RAMAN AND HIS EFFECT

Vignettes in Physics
A Series by G. Venkataraman

Published
The Many Phases of Matter
Why Are Things the Way They Are?
Bose and His Statistics
Chandrasekhar and His Limit
A Hot Story
At the Speed of Light
The Quantum Revolution (3 vols.)
1. The Breakthrough
2. QED: The Jewel of Physics
3. What is Reality?
Raman and His Effect
Bhabha and His Magnificent Obsessions
Saha and His Formula

Forthcoming
The Big and The Small

Vignettes in Physics

RAMAN AND HIS EFFECT

G. Venkataraman

Universities Press

RAMAN AND HIS EFFECT

UNIVERSITIES PRESS (INDIA) PRIVATE LIMITED

Registered Office
3-6-747/1/A & 3-6-754/1, Himayatnagar, Hyderabad 500 029, Telangana, India
info@universitiespress.com; www.universitiespress.com

Distributed by
Orient Blackswan Private Limited

Registered Office
3-6-752 Himayatnagar, Hyderabad 500 029, Telangana, India

Other Offices
Bengaluru / Chennai / Guwahati / Hyderabad / Kolkata
Mumbai / New Delhi / Noida / Patna

First published 1995
Reprinted 1997, 2001, 2004, 2005, 2006, 2008, 2009, 2011, 2014, 2015, 2018, 2024

502032

ISBN 978-81-7371-008-7

Typeset by
Access Center, Secunderabad 500 003

Printed in India by
B B Press, Tronica City, Ghaziabad, U.P. 201103

Published by
Universities Press (India) Private Limited
3-6-747/1/A & 3-6-754/1, Himayatnagar
Hyderabad 500 029, Telangana, India

Contents

Preface

To the adult reader

This book and others in this series written by me are inspired by the memory of my son Suresh who left this world soon after completing school. Suresh and I often used to discuss physics. It was then that I introduced him to the celebrated *Feynman Lectures*.

Hans Bethe has described Feynman as the most original scientist of this century. To that perhaps may be added the statement that Feynman was also the most scintillating teacher of physics in this century.

The Feynman Lectures are great but they are at the textbook level and meant for serious reading. Moreover, they are a bit expensive, at least for the average Indian student. It seemed to me that there was scope for small books on diverse topics in physics which would stimulate interest, making at least some of our young students take up later a serious study of physics and reach for the Feynman as well as the Landau classics.

Small books inevitably remind me of Gamow's famous volumes. They were wonderful, and stimulated me to no small extent. Times have changed, physics has grown and we clearly need other books, though written in the same spirit.

In attempting these volumes, I have chosen a style of my own. I have come across many books on popular science where elaborate sentences often tend to obscure the scientific essence. I have therefore opted for simple English, and I don't make any apologies for it. If a simple style was good enough for the great Enrico Fermi, it is also good enough for me. I have also employed at times a chatty style. This is deliberate. Feynman uses this with consummate skill, and I have decided to follow in his footsteps (whether I have succeeded or not, is for readers to say). This book is meant to be read for fun and excitement. It is a book you can even lie down in bed and read, without going to sleep I hope!

Naturally I have some basic objectives, the most important of which is to stimulate the curiosity of the reader. Here and there the reader may fail to grasp some details, and in fact I have deliberately pitched things a bit high on occasions. But if the reader is able to experience at least in some small measure the *excitement* of science, then my purpose would have been achieved. Apart from excitement, I have also tried to convey that although we might draw boundaries and try to compartmentalise Nature into different subjects, she herself knows no such boundaries. So we can always start anywhere, take a random walk and catch a good glimpse of Nature's glory. Where she is concerned, all topics are "fashionable". There is today an unnecessary polarisation of the young towards subjects that are supposed to be fashionable. To my mind this is unhealthy, and I have tried to counter it.

This series is essentially meant for the curious. With humility, I would like to regard it as some sort of a "Junior Feynman Series", if one might call it that. With much love, and sadness, it is dedicated to the memory of Suresh who inspired it.

To the young reader

Chandrasekara Venkata Raman is unquestionably the best-known scientist of modern India. Better known by his shortened name C.V. Raman, he is famous for the discovery of the Raman effect. To date, he is the only Indian to have won the Nobel Prize for contributions to science, *made from within the country*. But Raman is much more than just a Prize winner, however important that might be. We need to remember Raman equally for his fighting spirit and the values he cherished. Why? Read on and find out!

Acknowledgements

I am grateful to the Indian Academy of Science for permission to reproduce several figures from *Journey into Light.* I am also thankful to Mrs. Naga Nirmala for typing assistance.

G. VENKATARAMAN

1 *The Young Raman*

This book is about an amazing man who wrote a new and exciting chapter in the history of this country. His full name is Chandrasekara Venkata Raman. The first name was derived from his father R. Chandrasekara Iyer whereas Venkataraman was the name given to him. In school, the name changed first to C. Venkata Raman and later to C.V. Raman, a name that remained. It is even said that Raman shortened his name so that discoveries he made would be named after him! (I know one scientist who shortened his name, and it was rumoured that he too entertained similar ambitions but alas, he has had little luck.)

Raman was born on 8th November, 1888 in a village near Tiruchirapalli (Trichinopoly in those days) in South India. Raman's parents were R. Chandrasekara Iyer and Parvathi Ammal. Raman was the second child of his parents; the first born was Subramanian, later known as C.S. Iyer. Of course the world knows him better as the father of S. Chandrasekhar, the renowned astrophysicist (see the companion volume *Chandrasekhar and His Limit*).

When Raman was four years old, his father moved to Vishakapatnam to take up a lecturer's job in the Mrs. A.V. Narasimha Rao College. There he taught physics, mathematics and physical geography. Chandrasekara Iyer was also a good athlete, and took active part in sports and physical culture.

Unlike his father, Raman was not physically strong (at least in those days). But what he lacked in physical strength, he more than made up with his mental abilities. He was always a student of the first rank and a voracious reader. He eagerly read the books in his father's collection many of which were to inspire him later on. One of these was a volume by the famous German scientist Helmholtz – see also Box 1.1.

Box 1.1 In later life, Raman once discussed the books that influenced him. He said:

> A list of all the volumes I had to study [as a student] would be

> of terrifying length... But out of this welter of subjects and books, can I pick out anything really to mould my mental and spiritual outlook and determine my chosen path in life? Yes, I can and shall mention three books.
>
> A purposeful life needs an axis or a hinge ... This axis has been, in my own case, ... a certain abstract idealism ... which I can trace to the reading of Edwin Arnold's great book *The Light of Asia*. I remember being powerfully moved by the story of Siddhartha's great renunciation ... It always seemed to me a surprising and regrettable fact that the profound teaching of the Buddha has not left a deeper impression in the life of our country of which he was the greatest son that ever lived.
>
> The next set of books that I have to mention are the most remarkable works of all time, namely, *The Elements of Euclid* ... The pages of Euclid are like the opening bars of the music of the grand opera of Nature's great drama. So to say, they lift the veil and show to our vision a glimpse of a vast world of natural knowledge awaiting study.

Earlier I made a reference to the influence that Helmholtz had on Raman. What does Raman himself have to say about it? Raman speaks of Helmholtz as "the supremest figure" of modern times and adds:

> It was my great good fortune, while I was still a student at college, to have possessed a copy of an English translation of his great work *The Sensations of Tone*. As is well known, this was one of Helmholtz's masterpieces... I gathered from it a variety of problems for research which were later to occupy my attention and keep me busy for many years.

At the age of eleven, Raman passed out from school. He then spent two years studying in the college where his father worked. And at the tender age of thirteen, he was off to study at the Presidency College, Madras, with a scholarship in his pocket.

Young Raman was quite unimpressive in appearance, and at Presidency, they had not seen such a young student before. As Raman himself wrote later:

> Indeed in the first English class I attended, Prof. E.H. Elliot addressing me asked if I really belonged to the Junior B.A. class, and I had to answer him in the affirmative. He then proceeded to inquire how old I was.

Soon everyone realised that Raman was their best student, with a flair for English as well as for science, physics in particular.

When Raman passed the B.A. examination (standing first of course and sweeping all the prizes), his teachers suggested that he go to England to prepare for the Indian Civil Service (ICS) examination. The ICS was the forerunner of today's Indian Administrative Service (IAS). It was considered even more prestigious than the IAS is at the present time. Very few made it, and most of those selected were Britishers. For an Indian to enter the ranks of the ICS was considered the pinnacle of achievement.

Before going to England for further studies, one had to appear for a medical examination. The Civil Surgeon of Madras who examined Raman declared that Raman was medically unfit to go to England. This was the only examination that Raman ever failed, and in later years Raman said about this medical officer, "I shall ever be grateful to this man!"

Raman was least perturbed that he could not go abroad. Instead he enrolled in the M.A. class, choosing physics as his subject. By now he had become a minor celebrity in college and his Professor gave him complete freedom to do as he liked. About this Raman later said:

> Professor Jones (Professor of Physics) believed in letting those who were capable of looking after themselves to do so, with the result that ... I enjoyed a measure of academic freedom that seems almost incredible ... During the whole of my two years' work for the M.A. degree, I remember attending only one lecture ...

Raman made good use of the freedom given to him, particularly by trying out new experiments not mentioned in textbooks. Thus it was that he wondered what happens when light is diffracted obliquely—see Box 1.2. Indeed, in this case he even managed to get his findings published as a short paper in the prestigious

Box 1.2 Consider light falling on a narrow slit as in figure (a). The light here is normally incident. As you know, the slit produces diffraction bands. The formation of these bands is readily explained using the wave theory of light, and the theory is found in college texts. Raman wondered what would happen when light is incident obliquely as in figure (b). This problem had not been discussed either in a book or even in any research article. Raman then applied his mind and by reasoning came to the conclusion that when the angle is very close to 90°, the diffraction bands are no longer symmetric as they are in the case of normal incidence illustrated in (a). He then performed some simple experiments to verify his conclusions, and reported his observations in a paper published in

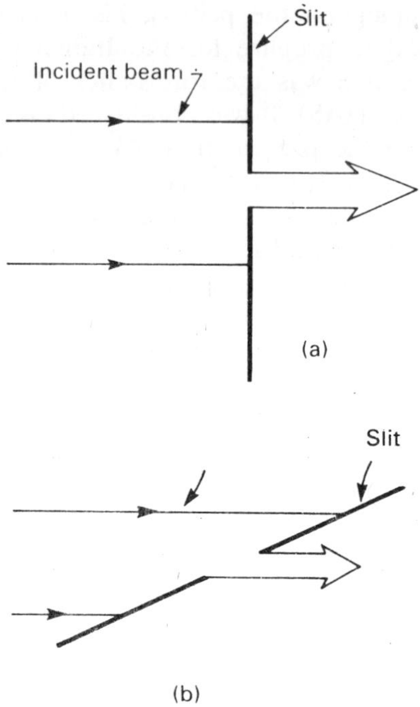

The Philosophical Magazine. There were still many questions that remained to be answered concerning oblique diffraction. These problems he pursued in Calcutta and they also resulted in publications.

British journal *The Philosophical Magazine*. He was the first student from Presidency College to publish a research paper. In this case it was all the more creditable because Raman had done this work independently without any assistance from his Professor. Another such paper resulting from his potterings in the college lab dealt with a new method of measuring the surface tension of liquids. This investigation was inspired by some remarks made by Lord Kelvin in his *Baltimore Lectures* (—see the companion volume *At the Speed of Light* for further comments on the Baltimore Lectures of Lord Kelvin).

Raman passed the M.A. examination in January 1907, standing first in the University and winning all the prizes available. His heart was clearly in science but opportunities for Indians for a research career in India did not exist. Going to England was also ruled out. Government service was the only available option, and so Raman took the Financial Civil Service (FCS) examination. The FCS was the forerunner of today's Indian Administrative and Audit Service (IA & AS), and Raman's elder brother C.S. Iyer was already a member of that service. Entrance to the FCS required not only a series of written examinations but also an interview. After the interview Raman declared, "I took one look at all the candidates who had assembled there and I knew I was going to stand first." And indeed he did!

Meanwhile, Raman got married to a young girl from Madurai named Lokasundari. In later years she was known as Lady Raman. And so it was that in the middle of 1907, while still in his teens, Raman went to Calcutta as Assistant Accountant General. His salary was Rs.400 per month, a fabulous amount in those days; it included the marriage allowance (see Box 1.3).

> **Box 1.3** In addition to their salary, officers of the Finance Department who were married, were entitled to an allowance of Rs.150 per month. This was a sizable addition to the salary, and Lady Raman often used to joke that Raman decided to get married in order to take advantage of this allowance!

Instead of going West, Raman went East. Instead of going to London, the capital of the British Empire he went to Calcutta, the capital of British India. And instead of entering science, he entered Government service. He was all set to get lost in obscurity and disappear into oblivion, but he did not.

In Calcutta, Raman used to go to office by tram. The tram used to pass a street called the Bow Bazaar Street. One day while riding to work, Raman saw a board with the words *Indian Association for the Cultivation of Science* (IACS) on it. His curiosity aroused, Raman got off at that spot on his way back and knocked on the door. The door was opened by a young man named Ashutosh Dey. It was a historic meeting because Dey, popularly known as Ashu Babu, was to become Raman's right-hand man for nearly quarter of a century.

I must now digress a bit to tell you something about the IACS. This Association was founded in 1876 by a man called Mahendra Lal Sircar. He was born in Calcutta in 1833. After completing school he joined the Calcutta Medical College and obtained the M.D. degree. But Mahendra Lal Sircar's interests ranged beyond medicine. Deeply nationalistic in outlook, he began to think hard about the country's problems. He then came to the conclusion that they can be solved only by the application of modern science. So, along with friends who thought like him, he founded a centre for scientific research on the pattern of the Royal Institution in London. By the way, the Royal Institution is the place where famous men like Michael Faraday and Sir James Dewar had done pioneering research. Sircar dreamt that something like that would happen in the Association he had founded. Unfortunately, nobody was keen on doing research at the Association and Sircar died in 1904 a disappointed man.

When Ashu Babu opened the door, Raman stepped in to find the place deserted. There was no sign of activity. Raman was intrigued and asked Dey what the Association was all about. Dey silently led Raman to Amrit Lal Sircar, the nephew of the late Mahendra Lal Sircar, who in turn wanted to know why Raman was interested. Raman then explained that although he was employed in Government service, he was keen on continuing research in physics; and to prove his credentials, he cited the papers he had published while studying at the Presidency College, Madras.

Amrit Lal Sircar could hardly believe his ears. All his life, the late Mahendra Lal was persuading people to use the facilities he had created and yet no one came. And here was a young man employed in Government service, offering to do research in his spare time, and without expecting a salary or even an honorarium.

It was too good to be true! Amrit Lal shed a silent tear in memory of his departed uncle and, welcoming Raman with open arms, exclaimed, "Where were you all these years?!" Raman was immediately handed over the keys of the building and given the freedom to come and go as he pleased.

Raman took full advantage of the offer and promptly moved into a neighbouring house. Professor Ramaseshan, a student of Raman, describes Raman's life in those days:

> 5.30 a.m. Raman goes to the Association. Returns at 9.45 a.m., bathes, gulps his food in haste and leaves for office, invariably by taxi [in those days there were no cars; by taxi is meant here a horse-drawn carriage] so that he might not be late. At 5 p.m., Raman goes directly to the Association on the way back from work. Home at 9.30 or 10 p.m. Sundays, whole day at the Association.

And so for ten years, Raman literally led a double life — at the office he was an efficient and diligent finance officer and the rest of the time he was a research physicist announcing his findings in leading scientific journals such as *Nature* and *The Philosophical Magazine* published in England, and in the *Physical Review* published in America.

Raman toiled alone, with Ashu Babu providing technical assistance By his researches, Raman made his name well-known to scientists in Europe and America even though he had not set foot out of the country. In Calcutta too Raman's reputation grew as he frequently delivered popular lectures on science. Raman's lectures were a treat not only because he had a good command of English but also because of his great sense of humour. And most important of all, his lectures were always accompanied by live demonstrations so that his audience could see at least some of the things he was lecturing about. At that time, popular lectures accompanied by demonstrations were quite common in England. Raman brought that tradition to India but alas, with him it also died.

What exactly did Raman investigate during this period of his research career? For some glimpses, turn to the next chapter.

2 *Science As A Pastime*

Raman stayed in Calcutta from 1907 to 1933. This period can be divided into two parts, 1907–1917 and 1917–1933. During the first part of this period, Raman was a Government officer doing scientific research in his spare time and during the latter, he was a Professor of Physics at the Calcutta University. In this chapter, I shall offer a glimpse of Raman's research during the first period.

A good bit of the work done during this period revolves around vibrations and acoustics. As examples, I shall consider Raman's work on musical instruments. I should mention in passing that much of Raman's interest in vibrations and acoustics was aroused by his earlier reading of the work of Helmholtz.

2.1 The peculiarity of the *ectara*

A string that is stretched and is therefore in tension, is capable of vibration. The basic vibrations are what are called the *natural vibrations* of the system. As opposed to this, the string can also be made to vibrate by an externally applied oscillating force. Such a vibration is called a *forced vibration*.

Free and forced vibrations are not peculiar to strings alone. The veteran Lord Rayleigh (see Box 2.1) had once investigated what

Box 2.1 Lord Rayleigh was one of the great pillars of classical physics. He was born in England in 1842. He passed the B.A. examination of Cambridge University in 1865 as a Senior Wrangler in Maths Tripos – a very difficult feat, which shows Rayleigh's class as well as promise. Soon after this, Rayleigh suffered an attack of rheumatic fever and was advised to spend some time in a warmer climate. So he went to Egypt, which at that time was under British influence. It was during this period that Rayleigh wrote his famous book on Sound. The remarkable thing is that he did not have access to a library while he wrote the book.

Refreshed by the trip, Rayleigh returned to England in 1873, built a small lab for himself and began to work. He could afford this. In 1879, the Cambridge University requested him to become the Cavendish

Professor, a position earlier occupied by Maxwell. Rayleigh agreed to the appointment, but only for a limited period of five years. From Cambridge he returned to his hometown.

Rayleigh was no doubt an outstanding mathematical physicist. But in addition, he was also a good experimenter, a fact that is hardly known. In fact, he was the first to isolate the rare gas argon. Prior to this experiment, argon was not known. Rayleigh found that the density of nitrogen isolated from the atmosphere was always greater than that of nitrogen prepared in the laboratory by chemical means. Long work eventually led to the discovery and the isolation of argon in 1895. For this work, he received the Nobel Prize.

Rayleigh's papers have been collected together and published. The collected works have inspired many a young student. For Raman too, it was some sort of a Bible.

happens when a tuning fork of frequency 256Hz is excited and mounted upon a resonance-box of frequency 512Hz – see Fig.2.1. He discovered that the box emitted a strong note at the frequency 512Hz, the primary tone of 256Hz being partly or at times even completely suppressed. Raman wondered whether there were other examples of systems emitting tones of frequency higher than their own.

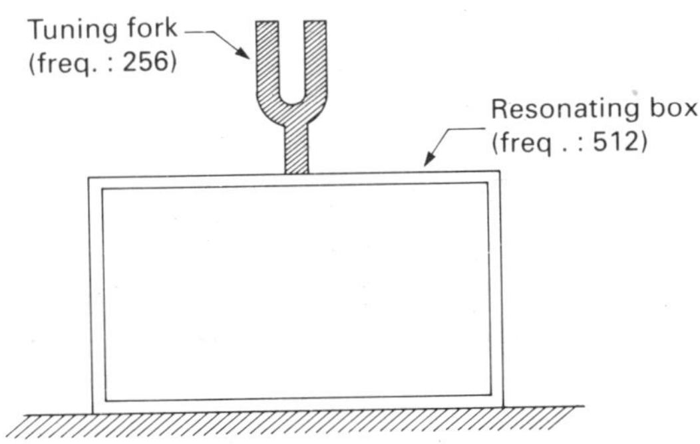

Fig.2.1 Schematic of the experiment of Lord Rayleigh that caught Raman's attention. The resonance frequency of the box is 512Hz. Therefore, when a fork of frequency 256Hz is placed on it, the box should not resonate; but it did, and this was Rayleigh's discovery.

The system he investigated is shown in Fig. 2.2. A metallic wire is attached perpendicularly to a board, passed horizontally over a

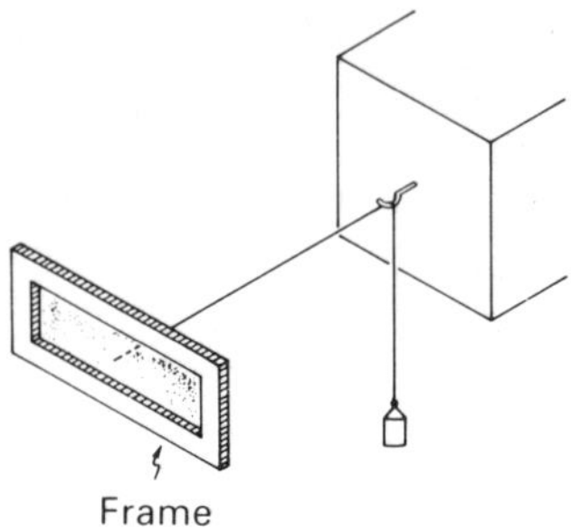

Fig.2.2 A wire in tension is attached to a board stretched on a frame. When the wire is plucked, the board vibrates at twice the frequency of the wire.

hook and weighted. With the wire in tension, it can either be plucked or bowed. When so disturbed and made to vibrate in one loop, a loud note was made by the sounding board. Raman showed by measurements that the frequency of the note emitted by the board was twice that of the frequency of the vibrating string, exactly as in Rayleigh's experiment. Raman then remarks:

> Strange as it may seem, there is in actual use in India, a musical instrument, a rather crude one, it is true, the working arrangements of which are, in essentials, the same [as in Fig. 2.2]. It is styled the *gopijantra* or oftener *ectara* [literally meaning single string], and is chiefly used, I find, by vocalists—those of the poorer sort—for striking key notes and marking time. The users of the instrument are apparently totally unaware of its unusual characteristics.

I am sure you must have seen sadhus and sometimes beggars going around with an instrument of the type shown in Fig. 2.3; this is the *ectara* that Raman is talking about. How does Raman explain his observation? He says that when the string is plucked, not only is the tension increased but so also the length. Thanks to this change of length, the wire tension varies periodically as $\cos 2pt$, where p is the frequency of the wire. Observe that the tension varies at *twice* the frequency of the wire. It is this periodic variation of the wire tension which, when communicated to the sounding board, causes the latter to vibrate at the frequency $2p$. Raman had an *ectara* specially constructed for his use and performed many experiments with it. Later he extended this work to discover what he called "remarkable resonances", but I shall not go into that.

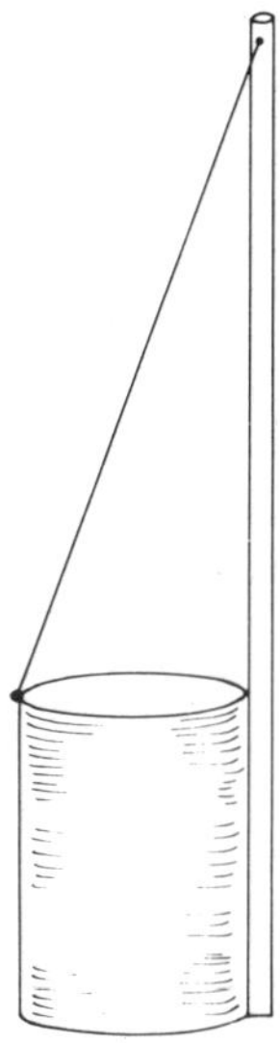

Fig.2.3 Schematic of the *ectara* used by sadhus.

2.2 Studies on the violin

Few know that Raman did some outstanding research on the physics of the violin. In fact, it still commands attention, and the method Raman introduced for characterising the quality of the instrument is used even today.

Let us start with the instrument itself. It is shown in Fig.2.4 together with the names of some of its parts. The four strings labelled G, D, A, E correspond (roughly) to the frequencies 196, 294, 440 and 659Hz. From the physics point of view there are three major components or subsystems, namely the string, the bridge which communicates the vibrational energy of the string to the box, and finally the box itself. The box absorbs energy from the string, vibrates and then radiates energy into the air – this is how sound is generated from the violin.

As you know, the violin is a musical instrument of Western origin. It has been hailed as the king of instruments. In the nineteenth century, Helmholtz undertook many studies on the violin and Raman had read about them while he was still a student. This was not all; Raman's father Chandrasekara Iyer played the violin well. Raman had often heard his father play the instrument and this probably

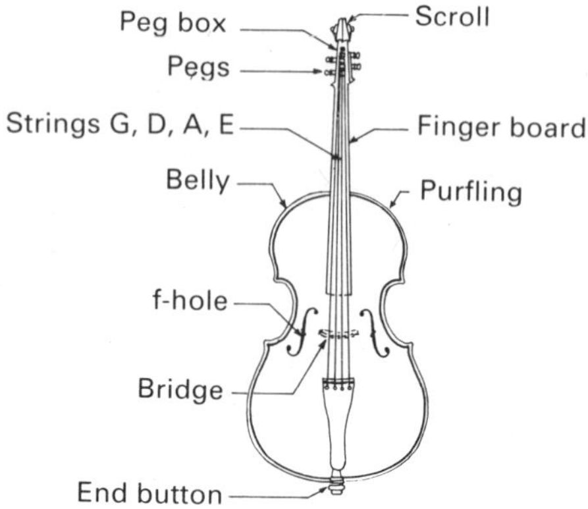

Fig.2.4 The violin and its various parts.

increased his curiosity about it. I should also mention that the violin came to India with the European colonialists. Soon the natives took to the instrument, although they adopted it to play Indian music and used a different playing stance.

Raman's studies on the violin were quite extensive and resulted in a whole book. Running to 158 pages, it is entitled *On the Mechanical Theory of Vibrations of Musical Instruments of the Violin Family with Experimental Verification of the Results, Part* I – quite a long title isn't it?! Obviously, I cannot discuss all of it here; instead, I shall give you just a flavour of Raman's monumental work. But before I do so, I should mention that Part II was never published. Indeed Raman often titled many of his papers as Part I but seldom did he produce Part II! He was always so full of ideas that he often drifted away to other investigations; the promised Part II thus remained a promise.

As you know, in the hands of an expert violinist, the instrument can be made to produce literally magical sounds. What is the control the violinist has? Among other things, he can choose the position on the string where the bow is applied. He can control the pressure of the bow, he can decide the speed of the bow, the stroke length and so on. While doing a *physics* experiment, one should be able to carefully control these various parameters. For this purpose, Raman designed a mechanical violin player which is shown in Fig. 2.5. It was assembled using lab junk and some parts bought from a cycle shop. An interesting feature of this is that in this machine, the bow remains fixed and it is the violin that moves.

Imagine a violinist playing the instrument that way! This is the first ever mechanical violin player designed. Since Raman's time, an automatic violin player has become an essential part of all violin research. Needless to say, the present-day gadget is very different, particularly in that the violin no longer moves.

Fig.2.5 The mechanical violin player devised by Raman.

Figure 2.6 shows one of the results obtained by Raman. What you see here is the bowing pressure that must be applied in order

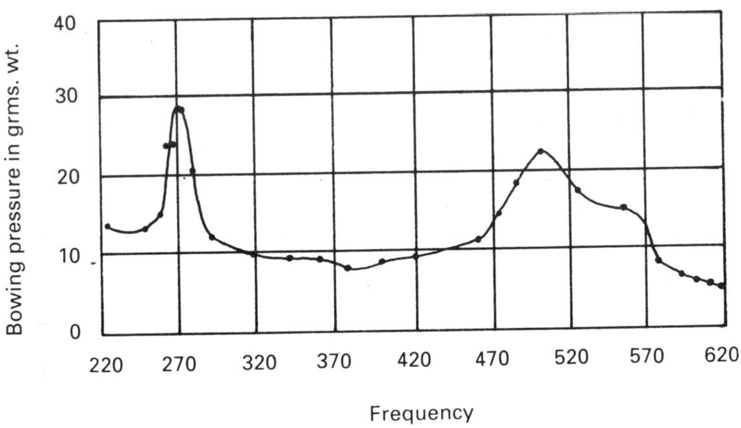

Fig.2.6 Frequency response of the violin as determined by Raman. The frequency was varied by controlling the string length with stops.

to make the violin resonate at various frequencies. Observe that at certain frequencies, more pressure has to be applied. The peaks are referred to as resonances. At the resonance frequencies, the bridge becomes very efficient in transferring the string energy to the box which is why more pressure is required.

Raman noted that the frequency response of a violin would say something about its quality. Such a frequency response curve is now called a *Raman curve*, and Fig. 2.7 shows how the study of Raman curves has improved with time.

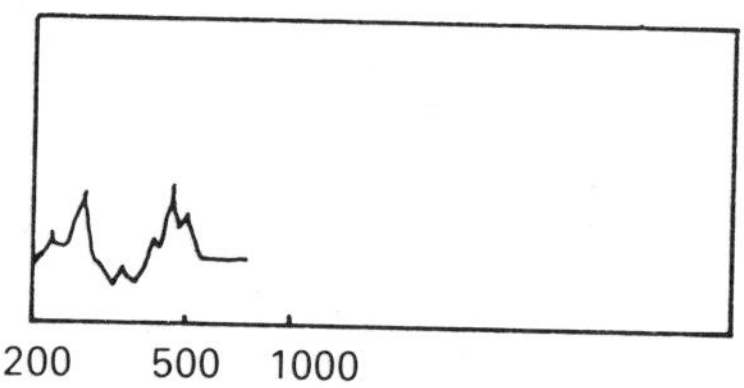

The original Raman curve

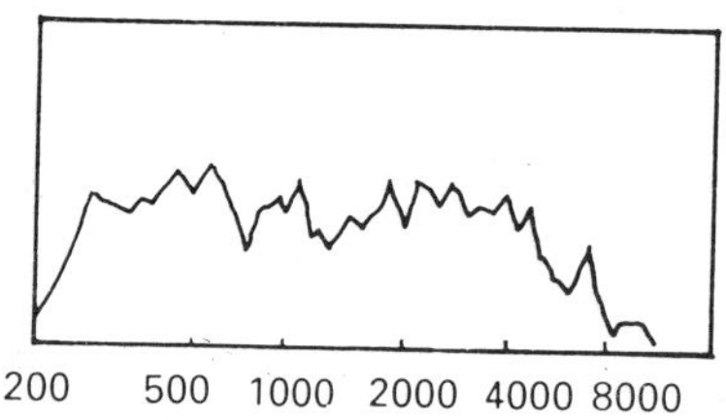

Response curve obtained by Saunders by means of the early method

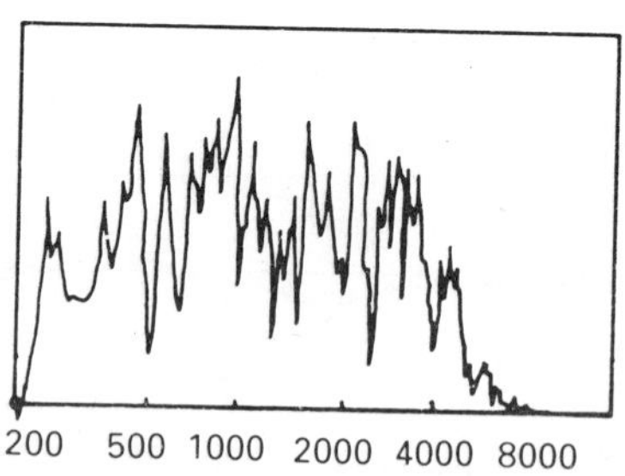

Response curve obtained by Saunders by means of a more improved method

Fig.2.7 This figure illustrates how the study of Raman curves has improved with time.

2.4 Raman's work on the *tambura* and the *veena*

I am sure you are familiar with the *tambura* and the *veena* – they are shown in Fig. 2.8. The *tambura* provides the drone or the *sruthi* for the musician. It has four metal strings stretched over the stem which ends in a big pumpkin-like "belly". The *veena* is similar but has frets. In both instruments, the strings are anchored to the belly after passing over a bridge. Figure 2.9 shows the bridges for the two instruments.

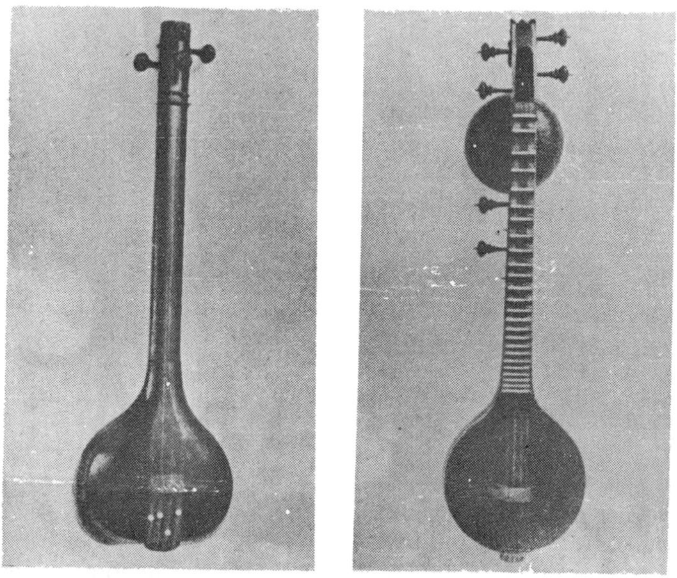

(a) (b)

Fig.2.8 The *tambura* (a) and the *veena* (b).

Raman discovered that the rich tonal quality of these two instruments is due to the peculiarity of the bridge. It is rounded and not sharp as in all Western musical instruments. Thus the forces exerted by the bridge on the resonating box are very different from what they would be, had the bridge been sharp. It is to this that Raman attributes the richness of the tone emanating from the *tambura* and the melody of the notes from the *veena*.

Unlike in the case of the violin, Raman did not make an exhaustive study of either the *tambura* or the *veena*. I believe that to

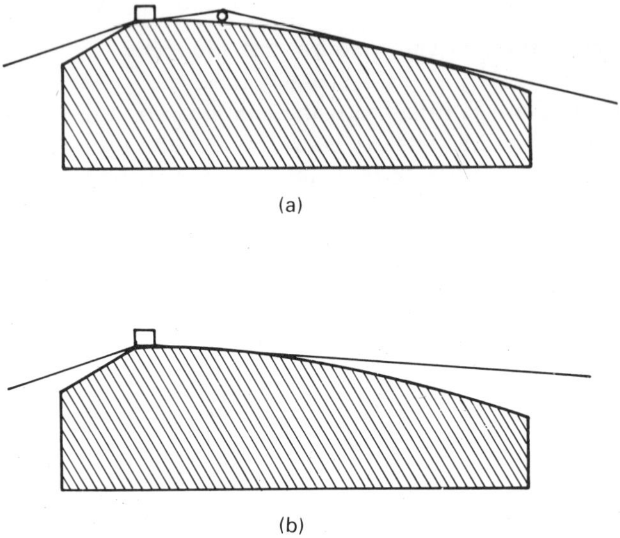

Fig.2.9 Bridges of (a) the *tambura* and (b) the *veena*.

this day, no such study has been undertaken, which is a pity. Here one must also stop and admire our ancients who intuitively designed instruments with remarkable musical qualities.

2.5 The *mridangam* and the *tabla*

Raman was fascinated by what he called the "remarkable appreciation of acoustical principles" on the part of our ancients. Their intuitive skills are particularly evident in the percussion instruments, the *mridangam* and the *tabla*—see Fig.2.10. Percussion instruments are also known in Western music; for instance, one has the drums. Comparing the two, Raman writes:

> All the instruments of percussion known to European science are essentially nonmusical and can only be tolerated in open air music or in large orchestras where a little noise more or less makes no difference. Indian musical instruments of percussion stand in an entirely different category. Times without number we have heard the best singers or performers on the flute or violin accompanied by the well-known indigenous musical drums, and the effect with a good instrument is always excellent.

Why? Raman supplied the answer.

(a)

(b)

Fig.2.10 (a) The *tabla* (b) the *mridangam*.

Consider first a uniform circular membrane. Lord Rayleigh has studied the different modes in which this membrane can vibrate. They are called the *normal modes of vibration*, and are illustrated in Fig. 2.11. The dashed lines that you see here are the *nodes*, i.e., the lines along which there is no movement. The way the membrane moves in time is shown in Fig. 2.12.

I now invite your attention to the frequencies of the different modes. The fundamental is the mode where the membrane goes up or down uniformly. If we take its frequency as 1, that of the next higher mode is 1.594. In the case of a vibrating string, this ratio would be 2, making the mode of higher frequency a true harmonic of the fundamental. If you examine the numbers given in Fig. 2.11 for the membrane frequencies, you will notice that none of them correspond to a harmonic.

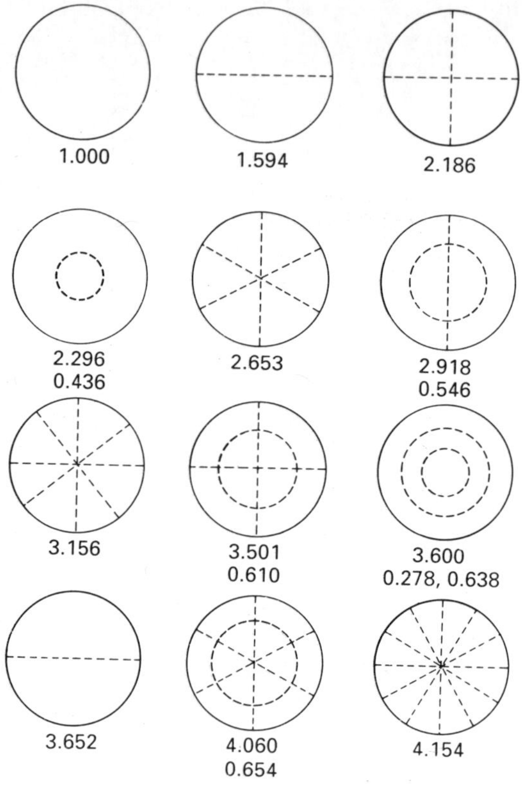

Fig.2.11 The important normal modes of vibration of a **uniform** circular membrane. The dashed lines show the nodal lines and nodal circles. The numbers in the upper line give the frequencies, taking the fundamental as unity. The numbers in the lower line give the radii of the nodal circles relative to that of the membrane.

Now when the membrane of a drum is struck, many modes would be simultaneously excited. However, since their frequencies are not in a harmonic ratio, they will not produce a pleasing sound; this is why the Western drum is a mere "noise producer".

What about the Indian drums? The membrane in this case is not a uniform one but a composite one whose thickness and density vary

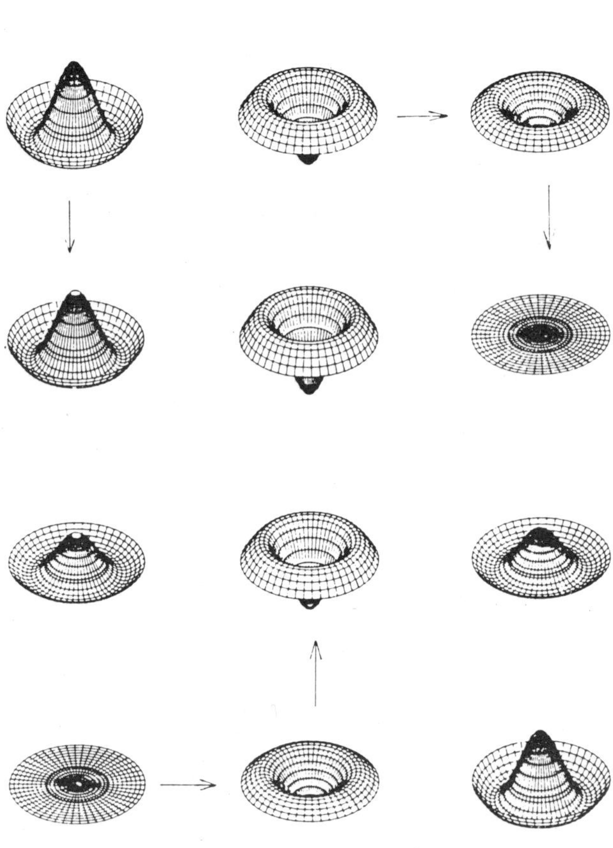

Fig.2.12 This shows membrane "snapshots" at various instants, as it vibrates according to one of the normal modes. In this case, there is one nodal circle. The pictures were generated on a computer.

somewhat as illustrated in Fig. 2.13. Mathematical analysis shows that, because of this nonlinear density distribution, the higher modes of vibration become harmonics, and that is what adds to their tonal quality. This was a beautiful though empirical discovery of our ancients, and Raman provided a scientific explanation. Next time you go to a music concert, pay attention to the drone of the *tambura* and the *swaras* emanating from the percussion instruments. If you are lucky, you might even hear the percussionist producing the notes corresponding to the seven swaras (*sa, ri, ga, ...*) by suitable striking of the drum.

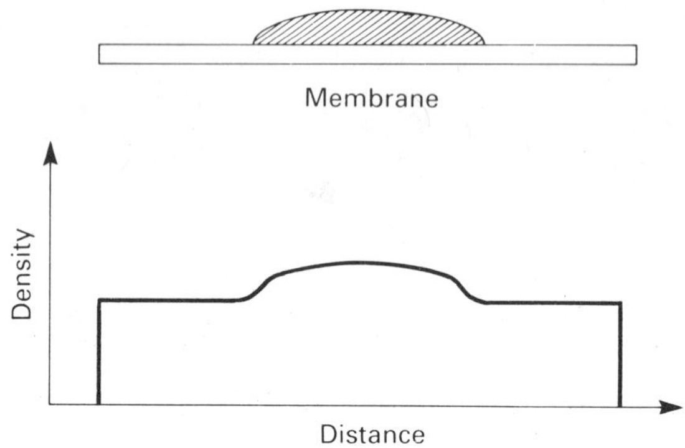

Fig.2.13 Schematic of the density variation of the membrane in an Indian drum.

Raman investigated the various nodal lines and circles by the simple technique of strewing a little fine-chalk powder on the membrane either before or immediately after striking the drumhead. The powder then gathers along the nodes in a clear-cut manner—see Fig. 2.14.

Raman investigated many other musical systems like the Stroh violin, the piano, the *jalatarung* and so on. He became such an authority on the subject of vibrations and the physics of musical instruments that he was invited to contribute an article to the physics encyclopaedia *Handbuch der Physik* published from Germany.

Raman's interest in vibrations and in the acoustics of musical instruments declined in 1920 because he had found a new love. More about that in a later chapter. But I should mention that among those interested in the physics of musical instruments, Raman is still a revered name. He was made an honorary member of the Catgut Acoustical Society (of America). In a message he once sent to them, he said:

> My studies on bowed string instruments represent a phase of my earliest activities as a man of science. They were mostly carried out between the years 1914 and 1918. My call to the Professorship at the Calcutta University in July 1917, and the intensification of my interests in optics inevitably called a halt to my further studies of the violin family of instruments.

And so we are now in the year 1917 when Raman quit his comfortable Government job to become a University Professor. For that story, turn to the next chapter.

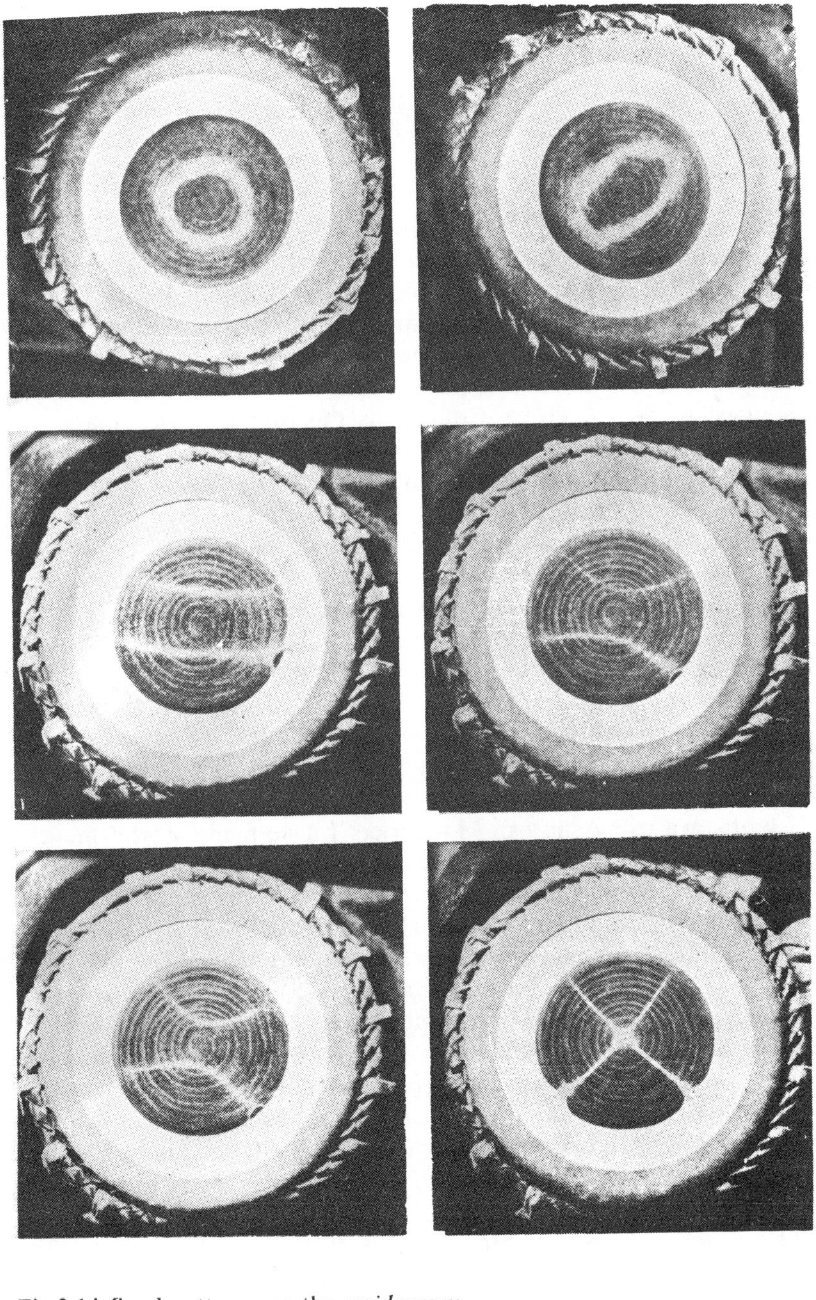

Fig.2.14 Sand patterns on the *mridangam*.

3 *Raman As Palit Professor*

Between 1907 and 1917, Raman was literally leading a double life. In the office, he was a Government official. Brilliant and efficient in work, he was much appreciated by his superior officers, most of whom were British. Everyone expected Raman to rise high in his profession, and some even thought that one day he would find a place in the Viceroy's Council as the Member for Finance. In today's parlance, that is like becoming the Finance Minister.

All this within the four walls of the office. Away from office, Raman was a brilliant physicist, researching problems of the type that had attracted giants like Rayleigh and Helmholtz. He had not yet travelled abroad but his name was known to scientists in other countries through his publications which, as I have already told you, appeared in leading scientific journals published from England and America. And to the people of Calcutta, Raman was not only a person who had made the dreams of late Mahendra Lal Sircar come true, but also one who made science come alive when he delivered popular lectures. In short, Raman was making waves as a scientist.

One person who was quietly watching Raman's emergence as a scientist was Sir Ashutosh Mookerjee. I have made a brief mention of him in the companion volumes *Bose and His Statistics* and *Saha and His Formula* but I must say something more about him, particularly here, as he played a crucial role in the life of Raman.

A brilliant scholar, Sir Ashutosh was trained in law. That was also his profession, and soon he was appointed a Judge of the Calcutta High Court. For an Indian to be appointed to the High Court in those distant days was a great honour. It also meant that Sir Ashutosh was exceedingly good in his profession.

Like Raman, Sir Ashutosh had a second love; in fact, as in the case of Raman, it was science. Sympathising with the aims of Mahendra Lal Sircar, Sir Ashutosh had worked with him in founding the IACS. Later, he spent a lot of time delivering lectures at the Association on topics in mathematical physics. It is no wonder therefore that Sir Ashutosh greatly admired Raman not only for enlivening the then sleepy Association but also for bringing it international fame.

In 1916, Sir Ashutosh Mookerjee was appointed the Vice-Chancellor of the Calcutta University. He immediately took steps to strengthen science teaching and research in the University. Until he took over, the University mainly functioned as an examining body and did not have a science department of its own. Students who wanted to study science had to join the Presidency College (like Madras, Calcutta also had a Presidency College; it was a Government College of course), study there and then appear for the examination conducted by the University.

Ashutosh Mookerjee collected funds and not only started a Science College (as a part of the University) but also managed to endow Professorships in Physics and Chemistry. Having done this, Sir Ashutosh was on the lookout for capable people who could be appointed to various positions. Thus it is that his eyes fell on Raman as a possible candidate for the University Professorship in Physics.

Around 1916–1917, Ashutosh Mookerjee approached Raman with the offer. At that time, Raman was drawing a salary of Rs.1100 per month which, considering the value of the rupee then, was a truly fabulous sum. As against this, the Palit Professorship which was being offered carried a salary of only Rs.600, minus all the perks that come with a prestigious Government job. Would Raman like to switch over? Raman was immediately interested. Would he not lose money? And what about power, prestige, status and above all the prospect of rising high in Governmental circles? Yes indeed, he would lose all these as well. As against such losses, he could, if he accepted the Professorship, devote *all* his time to physics. Added attraction, he could have students to work with him, which also meant that he could expand his research interests.

Raman was definitely interested. He asked the Government to give him two years leave so that he could try out being a Professor. If there were problems in adjusting, he would come back to his old job. In short, he wanted leave (for academic purposes), with a lien on his job. When Raman applied for leave, his superior officers were shocked. They told him: "Look, you must be crazy. Why do you want to spoil a brilliant (Government) career? In the not-too-distant future, you have excellent chances of becoming the Member for Finance. How many people would have such a chance? And you want to throw it all away? For what? Doing physics? But you are already doing physics research. Why don't you just continue as at present and get the best of both worlds?"

Raman would not listen. "Will you give me leave or will you not?" – that was his only response. And when the Government refused leave, Raman chucked his cushy job without batting an eyelid. Even today, few would take such a daring step but Raman did it three-quarters of a century ago when the British Empire was at its peak and few would dream of giving up so much, especially at the risk of annoying a powerful Government.

Sir Ashutosh was not only pleased but also deeply touched. He observed:

> I should fail in my duty if I were to restrain myself in my expression of genuine admiration I feel for the courage and spirit of self-sacrifice with which Mr. Raman has decided to exchange a lucrative official appointment with attractive prospects, for a University Professorship, which, I regret to say, does not carry even liberal emoluments. This one instance encourages me to entertain the hope that there will be no lack of seekers after truth in the Temple of Knowledge which it is our ambition to erect.

That in brief is how Raman became the Palit Professor of Physics at the Calcutta University. There was a little problem though which first had to be overcome. According to one of the conditions laid down by the Trustees who gave the money for the Palit Professorship, the person appointed must have had "foreign" experience. Raman, however, could claim no such experience having upto that time, never stepped outside the country. A quiet suggestion was made that he should go abroad for a few months and pick up the required "foreign experience". Raman brusquely dismissed the suggestion. In his typical fashion he declared that he was ready to train foreigners! Indeed, with his international reputation, he was. Ways were then found to exempt Raman from this legal requirement, and after that he could take up the appointment.

The Palit Professor was not required to teach students on a regular basis. His job was mainly to manage the lab in the Science College and more importantly, to guide students in research. This suited Raman admirably, and he began to operate on two fronts, i.e., at the University and at the Association. Physics was no longer a part-time affair but a full-time occupation.

Though he was not required to do so, Raman took to teaching also because he loved it. L.A. Ramdas who studied under Raman in Calcutta (and later rose to a high position in the India Meteorological Department) writes:

> Prof. Raman took 'Electricity and Magnetism' in the year 1920–21 and 'Physical Optics' in 1921–22 ... students felt that they were

> indeed listening to a type of inspired teaching to which was brought all the original flavour and excitement of the great giants of the past ... We shared with him much of the excitement and superb thrill that Benjamin Franklin, Oersted, Arago, Gauss, Faraday, Maxwell, Hertz, Lord Kelvin and many others must have felt while they were making their actual discoveries ...

Equally important, Raman induced the students to learn from the original books and papers of the founding fathers of each subject.

Thanks to the University job, Raman now began to attract research students. Such was his reputation that they came to him not only from Calcutta but also from far off places. From Trivandrum came K.R. Ramanathan, from Madras came K.S. Krishnan and from Andhra came Suri Bhagavantam (see Box 3.1, 3.2 and 3.3). At the Science College the students attended classes whereas at the Association they did research. Raman himself shuttled between these two places, and for all practical purposes, the Association became the research arm of the University.

Box 3.1 K.R. Ramanathan was born in Kerala in 1893. He studied for the Bachelor's degree in the Victoria College in Palghat and later joined the Presidency College, Madras to study for the M.A. degree. One of the examiners who came for the practicals, Professor Stephenson of the Maharajah's College of Science, Thiruvananthapuram (in those days it was Trivandrum) was so impressed that he offered the young scholar the post of Demonstrator on the spot! Ramanathan worked in Trivandrum for seven years, during which period he also served in the local observatory. This experience enabled him to write a paper on tropical thunderstorms.

Attracted by Raman's fame, Ramanathan left for Calcutta where he earned his doctorate. He then went to Rangoon to serve as a lecturer but he rushed to Calcutta on every conceivable opportunity to work in the Association. In 1925, he joined the India Meteorological Department where he served until retirement at the age of 55. As a meteorologist, Ramanathan made many valuable contributions which earned for him international reputation. Thanks to this, he received invitations from abroad, and it is interesting that Ramanathan went abroad for the first time *after retirement*!

From the Met office, Ramanathan went as Director to the Physical Research Laboratory in Ahmedabad, which had been founded by Vikram Sarabhai. In 1966 he stepped down from the Director's post but continued to be active in research at the lab till the very end – a touch of Raman. During his life Ramanathan received many honours. He passed away in 1985.

Box 3.2 K.S. Krishnan is one of the eminent physicists produced by the country in the pre-war era. He was born in Watrap, Tamil Nadu in 1898, and received his education in Madurai and Madras. After obtaining the Master's degree, he worked as Demonstrator for a brief period. Attracted by Raman's fame, he went to Calcutta to study and work under Raman. You will hear more about him and his close association with Raman during the days prior to the discovery of the Raman effect.

After obtaining his doctorate degree, he spent a few years at the Dacca University where Satyen Bose was also teaching. In Dacca, Krishnan switched fields and started working on diamagnetism. From Dacca, Krishnan went back to Calcutta. For a brief period, he was the Mahendra Lal Sircar Professor at the Association (this was after Raman had left). From Calcutta, Krishnan went to Allahabad and from there to Delhi as the Founder–Director of the National Physical Laboratory (NPL).

Krishnan was a great scholar, and besides physics he enjoyed Tamil, Sanskrit and English literature. He was widely respected. Jawaharlal Nehru was an admirer of Krishnan, and he once said that he would rather be the Director of NPL than Prime Minister! For his scientific achievements, Krishnan was elected a Fellow of the Royal Society and a Member of the U.S. National Academy of Sciences. He passed away in 1961.

Box 3.3 Suri Bhagavantam was born in 1909 in Andhra Pradesh. Like Ramanathan and Krishnan, he was attracted by Raman's fame and headed for Calcutta. He arrived in Raman's lab at the right time, i.e., when the Raman effect was about to be discovered.

In a later chapter, you will read about Raman's migration to Bangalore to take up appoinment as the Director of the Indian Institute of Science. Raman tried to get Bhagavantam appointed as his scientific assistant but failed. Bhagavantam then went to the Andhra University, where he spent fifteen years. There he wrote two books which became famous. These are: *Scattering of Light and the Raman Effect*, and *The Theory of Groups and its Applications to Physical Problems* (which he coauthored with T. Venkatarayudu). Later Bhagavantam served as the Vice-Chancellor of Osmania University in Hyderabad, as the Director of the Indian Institute of Science and as the Scientific Adviser to the Defence Minister. He maintained a strong interest in research till he retired. He was an excellent teacher. He passed away in 1989.

Raman's life continued to be hectic and about it Ramdas tells us:

> Often he [Raman] used to work [at IACS] far into the night and when exhausted, sleep on a table until the astonished Ashu Babu awakened him next morning! On most mornings he would come in informal dress to the Association and carry on experiments till 9.30 a.m., when he would remember his lecture engagement at the University ... A quick return home, a few strategic sweeps of his razor for shaving, a quick bath, a hurried dressing along with the inevitable Madras turban, and he would shout at the top of his powerful voice, "Ashu Babu – taxi", which would evoke an equally loud response from Ashu Babu! Snatching his breakfast in a minute or two he would rush into the Association with a sheaf of his lecture notes for the day, jump into the waiting taxi with his laboratory attendant Shivanandan, and after a hectic drive reach the Science College just in time to start his lecture.

In 1919, Amrit Lal Sircar passed away and Raman became the Honorary Secretary of the Association. Thanks to the influx of students, the Association now hummed with activity at all hours. Work also branched into areas other than acoustics – in particular, optics began to receive increasing attention.

After graduating from college Raman became a Government officer and from the Government job he straightaway stepped into a University Professorship on the strength of his research output. Not for him the standard route of first winning a doctorate degree etc. In 1921, the Calcutta University conferred on him a honorary doctorate degree, but to his students he was always "Professor"; occasionally, it was Prof. Raman.

Nineteen twenty-one was also the year Raman went abroad for the first time. The official excuse was to attend the University Congress at Oxford where he represented the Calcutta University. For Raman, the trip was a great experience in more ways than one. Firstly, he got an opportunity to meet face to face many of the eminent physicists of the day. Secondly, he got a chance to do some research in England. This included an investigation of the acoustic properties of the central gallery in the St. Paul's Cathedral in London (the whispering gallery phenomenon). Later this triggered him to investigate similar domed structures in India. And finally there was the memorable return voyage which got him thinking about a new line of research that eventually lead him to fame; that story comes in the next chapter.

Here I must interrupt to narrate a story that the famous physicist Chandrasekhar told me.

In those days, going abroad was a very rare event and whenever anyone did, others were immensely curious to learn of the traveller's experiences. Soon after he returned from his first ever foreign trip, Raman visited his elder brother C.S. Iyer in Madras. A small crowd collected in Iyer's house to hear from Raman about his experiences. Being a lively narrator, Raman kept his audience roaring with laughter. And then suddenly a young boy in the crowd asked, "Did you not find it embarrassing to move around London in a turban?" Raman looked at this questioner and said:

> Young man, I will tell you about a little incident that happened while I was in London. One evening, I went to the Royal Institution to hear a lecture by Lord Rutherford. I arrived a little late and by that time the lecture had started. And so I quietly slipped into one of the back rows and sat there. Suddenly, Lord Rutherford looked at me and said, "Professor Raman, why are you sitting there all alone in the back row? Come up here to the front." I then went and sat in the front row with all the famous British scientists. After the lecture I went to Rutherford and asked him, "Professor Rutherford, how did you recognise me? This is my first trip to England, and we have never met so far." Rutherford replied, "Well, I have read your papers and when I saw a person in the audience wearing a Madrasi turban, I knew it must be you." So young man, now you tell me what is wrong with a Madrasi turban?

In 1924, Raman was elected a Fellow of the Royal Society. This was a signal honour and a much deserved recognition of the pioneering work he had done over a decade, almost single-handed. There was a party to felicitate Raman during which Sir Ashutosh asked, "What next, Raman?" to which Raman replied without any hesitation, "Why sir, the Nobel Prize of course!" Sounds like a brash answer brimming with overconfidence but that was Raman.

Raman waited thirty-two years before making his first foreign trip and then all of a sudden, trips came one after another. In 1924, he went across the Atlantic. First stop was Canada, where he went as the guest of the British Association for the Advancement of Science. This Association was holding a meeting in Toronto, Canada, and Raman was invited to lead the discussion on light scattering, a subject he had become interested in, following his return from England in 1921. He then crossed over to the United

States to spend three months at the California Institute of Technology. This was on an invitation from Robert Millikan, famous for his experiments on the charge of the electron. The Visiting Professorship offered to Raman was indeed very prestigious because among those who had held it earlier were Einstein, Lorentz and Sommerfeld. Years later Raman returned the compliment by playing host to Millikan in Bangalore.

On his way back from America, Raman stopped over at many places in Europe. He returned to India in March 1925. A few months later he was off again, this time to the Soviet Union (now defunct). He visited various scientific laboratories in Russia. During this visit, Raman conversed with Russian scientists mainly in German. In those days, Germany was very strong in physics and most physicists made it a point to learn to read German, if not speak it as well. Raman returned from Russia via Berlin, Paris, Geneva and Rome.

After this trip, Raman laid off foreign travels for several years, concentrating intensely on research. And it was during this period that the Raman effect was discovered. That story is reserved for the next chapter.

In 1928, Arnold Sommerfeld (see Box 3.4) visited India. Actually, Sommerfeld was going to America and the normal route from Germany to America was across the Atlantic. But Sommerfeld had heard about India, he had met Saha (see Box 3.5) in Europe and he was also familiar with Raman's work on optics. He therefore decided to go to America via the East, i.e., via India and Japan.

Box 3.4 Arnold Sommerfeld is one of the great names in classical physics. He was born in Germany in 1868. He was always interested in mathematics but studied many other subjects as well, including economics. He also took a great deal of interest in fencing! His instructor in mathematics was the great David Hilbert. For a while, Sommerfeld served in the army. Towards the end of the nineteenth century, Sommerfeld moved to Gottingen where he worked with the great mathematician Felix Klein. Here it was that he did his first major piece of work which was to work out an exact solution to a difficult diffraction problem. This solution attracted attention from all over, including that of the well-known French mathematician Henri Poincare. Sommerfeld was a great teacher and not only produced outstanding students like Pauli, Bethe, Heisenberg and Debye (all of whom won the Nobel Prize), but also wrote a series of books on theoretical physics. In those days, this was the series to learn from (the currently popular series is due to Landau). Physicists always think of Sommerfeld as an outstanding theoretical physicist, which of course he was, but few seem to know that he has done much work in engineering also. The vibration of bridges, problems of locomotive construction and the lubrication of machines are some examples. Sommerfeld died in 1951 in an automobile accident.

Box 3.5 Meghnad Saha was born in a village near Dacca in 1893. Today Dacca is the capital of Bangladesh but in those days, it was a part of British India. Saha belonged to a family of shopkeepers. They were not very rich. Young Saha was keen on studies and won a scholarship to study in school. In 1905, when the Governor of Bengal came to Dacca, Saha participated in a demonstration against him (remember, we were then under British rule and frequently there were agitations against the rulers). Because he joined the protest march, the scholarship was discontinued and Saha was now in trouble. He faced many difficulties but somehow managed to complete school education. In 1911, he joined the Presidency College in Calcutta. After college Saha tried for a government job but was disqualified because he had participated in the nationalist movement. Sir Ashutosh Mookerjee saved him by offering him a lecturer's job in the University College of Science. Here Saha began his research career and did his famous work on the *Saha Ionization Formula*, now a cornerstone of astrophysics. In 1923, Saha became Professor of Physics in Allahabad University. In 1937, he returned to Calcutta to become the Palit Professor of Physics (Raman had vacated that post earlier). Saha was also closely connected with the Association and was responsible for its movement from the original Bow Bazaar site to the modern campus at Jadhavpur. Saha also founded the *Institute of Nuclear Physics* which now appropriately has been named after him. In later years, Saha became involved in politics. He was a member of the Lok Sabha for many years. He passed away in 1956. For more about Saha, see *Saha and His Formula*.

Sommerfeld made careful plans for his visit to India. In this he was assisted by Saha. He had many lecturing engagements in India, one of which was at the Presidency College, Madras. A keen young member of that audience was Subrahmanyan Chandrasekhar and in the companion volume *Chandrasekhar and His Limit*, I have told you the story of the impact made by this lecture on Chandra.

On October 4, Sommerfeld arrived in Calcutta, the "main attraction" of his tour of India. Raman had made elaborate plans for Sommerfeld's stay.

Sommerfeld gave a series of lectures on the new subject of wave mechanics. He also got a first-hand account of the various experiments performed by Raman and his students in optics. And of course there was the demonstration of the famous Raman effect by the discoverer himself. Sommerfeld felt pleased about it all and in his diary he wrote: "Everything in the institute [IACS] is very good but the bathroom is terrible!"

In 1930, Raman received the Nobel Prize for his discovery of the Raman effect and he was also conferred the knighthood by the British Government which meant that his name now changed to Sir C.V. Raman. Later in life, Raman was simply called Sir C.V.

Around the same time, Raman was appointed a member of the Governing Council of the Indian Institute of Science in Bangalore. You will hear more about this institute in a later chapter. In those days, the head of the institute was always a Britisher. In the early thirties, the position of Director became vacant. It was the custom to approach eminent scientists to recommend a suitable name. On this occasion Rutherford was one of those so approached and he replied, "Why are you looking for a Director from England when you have an excellent person like Raman in India?"

And so it was, in early 1933, Raman received an offer to become the Director of this prestigious institute in Bangalore.

Bangalore offered many attractions but at the same time, Raman did not want to break off completely with Calcutta. He was hoping to found a new school of physics in Bangalore and build strong ties between it and the school of physics he had earlier created in Calcutta. Unfortunately, that was not to be.

Over the years, Raman had made many enemies in Calcutta. Some were quite jealous of Raman and hated his success while others had complaints against what they claimed was Raman's bossy style of functioning. Thanks to such opposition, Raman found one day that he was voted out of the office of the Honorary Secretary.

Raman was deeply hurt. He felt humiliated and wronged beyond reason. In days when the Association was but a mere building, he had built up its reputation almost single-handed by slogging in his spare time. He had trained students who had later become distinguished. And with his sensational discovery, he had placed Calcutta on the scientific map of the world. For all this, he was being paid by being thrown out of the office he had held with distinction for so many years after Amrit Lal Sircar had passed away.

It is difficult to pass judgement on these events which occurred so many decades ago. Let me just say that there were faults on both sides – Raman should have realised that over the years feelings against him were developing; he did little to counter them or prevent them from growing. For their part, those who were against Raman need not have gone to the extent they did, because quite unknowingly, they inflicted a deep wound on Raman whose scar remained all his life.

The upshot of it all was that Raman decided to leave Calcutta for good, instead of shuttling between Calcutta and Bangalore as he had originally intended. And when Raman finally left, Dr.L.L. Fermor, a member of the Geological Survey of India and a leading British scientist, said:

> Calcutta's loss will be Bangalore's gain. At present Calcutta may be regarded as the centre of scientific research in India; but with the transference to Bangalore of one of her leading investigators she will have to guard her laurels.

Prophetic words, because the seeds that Raman sowed in Bangalore have gone a long way to make that city into the scientific capital of India.

4 *Why Is The Sea Blue?*

4.1 Introduction

Raman's interest in optics became evident while he was in college. In fact his very first paper was on optics. This interest continued even while he was doing research at the Association. And later when he became the Palit Professor, he was able to assign various problems in optics to his students. It is these studies in optics done by Raman and his students that first caught the attention of Sommerfeld.

By optics, we normally mean the behaviour of light rays or light waves as they pass through a medium. Among other things, light can be reflected, or diffracted as it goes past an edge, or interfere when passing through many slits, etc. In 1919, Raman wondered about the connection between the phenomenon of diffraction and molecular scattering. His interest in this topic was aroused by an experiment that Keen and Porter had earlier performed.

Keen and Porter passed a beam of white light through a tank containing a solution. Certain chemicals were then added to this solution so that it gradually changed from a clear liquid into a turbid one. This was due to the production of particles which then remained suspended in the liquid. Naturally, the intensity of the transmitted light decreased as the turbidity increased, and at one stage the light was almost cut off. But now there was a surprise. Keen and Porter found that with the further passage of time, light began to be transmitted again. Not only that; the colour of the transmitted light went through a series of changes – indigo, blue, blue-green, greenish-yellow, and finally white. Why did such a thing happen? Keen and Porter did not have any explanation. Lord Rayleigh was able to explain the initial decrease in intensity of the transmitted light but had no answer for the strange appearance of colours later. Raman was able to explain all this. He did this by considering first the way the light wave is diffracted by the individual particles and then how these different diffracted waves combine together to producc an overall effect. This led him to wonder: What if the diffraction is not by a suspended particle but by a molecule?

4.2 On board the S.S. Narkunda

The year 1921. Raman is returning to India after his first ever visit abroad. No air planes in those days, and he has to travel by steamer. He boards the *S.S. Narkunda* at Southampton, England in early September. The journey to Bombay would take about two weeks, first around Spain, then past Gibralter into the Mediterranean sea, then via the Suez canal into the Red sea and finally into the Arabian sea for the home stretch to Bombay. To keep the passengers from getting bored during long journeys like this, the ship captain usually arranges games, parties, etc. Raman is least interested in such entertainment. Instead, he spends all his time on the deck, watching the sea. He is fascinated by the deep blue colour of the Mediterranean and begins to wonder why the sea is blue. Earlier, Lord Rayleigh who had successfully explained the blue colour of the sky (see Chapter 5) had declared, "The much-admired dark blue of the deep sea has nothing to do with the colour of water but is simply the blue of the sky seen in reflection." In short, the sea is blue because it is merely reflecting the blue sky – this was Lord Rayleigh's explanation.

Raman recalls Lord Rayleigh's explanation but is not convinced by it. He has with him some simple optical tools like a prism, a pocket telescope and even a pocket spectroscope. (Like a doctor who always carries his stethoscope, Raman always carried these gadgets.) While other passengers are frolicking, Raman spends hours observing the sea and examining it with the various optical instruments and gadgets he has with him. September 18, the ship reaches Aden. Passengers get out to feel Mother Earth, do some sight-seeing and buy curios in the souvenir shops. Raman instead rushes to mail a paper to *Nature* containing his observations made while crossing the Mediterranean.

He is not through yet. Rather, the Aden paper was just the beginning. As *Narkunda* races through the Arabian sea and as passengers are busy saying goodbye to each other and packing in preparation for disembarkation, Raman is busy completing yet another paper to *Nature*! This he mails at Bombay Harbour.

4.3 Back in Calcutta

From Bombay, Raman dashes off to Calcutta and promptly immerses himself in intense experiments. He is now convinced that (i) light can be scattered by the molecules of water just as it can be by the molecules of air, and (ii) that the blue colour of the sea is due to such molecular scattering just as the blue of the sky is. The thoughts and the experimental findings are gathered into a long paper which is then communicated to the Royal Society by Dr. Gilbert Walker (see Box 4.1). In it Raman says: "In this phenomenon, as in the parallel case of the colour of the sky, *molecular diffraction* determines the observed luminosity and in great measure also its colour."

> **Box 4.1** Gilbert Walker, F.R.S., was the Director-General of Observatories. He communicated many of Raman's early papers to the Royal Society. To Walker also belongs the distinction of having helped the mathematical genius Srinivasa Ramanujan at a crucial stage. As DGO, Walker had to visit observatories and inspect the measuring instruments. During an inspection trip to the Madras harbour, the Chairman of the Madras Port Trust, Sir Francis Spring asked Walker to take a look at the mathematical scribblings of an apparently eccentric clerk. Walker quickly recognised that the dreamy clerk had the most unusual mathematical abilities. He then helped him get a scholarship in the Madras University. Later Ramanujan went to England, worked with Hardy, and achieved undying fame. That part of the story is well-known.

Raman thus proved that the sea is blue because the molecules of water scatter light just the same way molecules of air do. Now light from the Sun is made up of all wavelengths and appears to us as white. Why then does the sky appear blue? That, said Lord Rayleigh, is because the light that reaches our eye is the scattered light (unless of course we see the Sun directly which we cannot because of its extreme brightness). Light of different frequencies is scattered with different efficiencies, the scattering varying as ν^4. In other words, the higher the frequency, the greater is the scattering. Since blue light has a much higher frequency than red, it is scattered much more efficiently; consequently the sky appears blue. If one goes into space, there is no atmosphere there and the sky appears black even if the Sun is shining. Perhaps you have seen photographs taken by astronauts. Incidentally, Rayleigh's discovery also enables us to understand why the railways use red light for signalling a stop. This is because red light can travel long distances without getting scattered. The engine driver can therefore see the signal from a sufficiently long distance and apply the necessary brakes. If blue light is used, the driver would not be able to see the signal until he comes close enough, and by then it might be too late. Getting back to Raman, it is only after completing the paper on scattering from water that he could even think of going south to visit his wife and child (who were there while Raman was away in England) and his brother C.S. Iyer in Madras. The anecdote which I narrated in the last chapter relates to this visit.

4.4 Molecular diffraction of light

Raman did not believe that his job was finished with the explanation he had just provided for the blue colour of the sea; in fact it was just the beginning. He let his thoughts wander further in an essay entitled *The Molecular Diffraction of Light* he wrote in 1922. People knew that light could be scattered by objects, just as it could be diffracted by slits, edges, spheres and the like. And now from the explanations given for the colour of the sky and the sea, it was clear that light was also being scattered by molecules. However, until then (1922), light scattering was mainly considered in terms of the classical wave theory of light. On the other hand, there was increasing evidence (see some of the companion volumes) that light energy was quantised. How does one consider the diffraction of these light quanta by the molecules? What effects arise in the process? Does anything new and not known before happen as a result of the quantum nature of light? Answers were not known then and they would not be for a few more years to come. But Raman boldly speculated on these issues in his essay which, by the way, was dedicated to Sir Ashutosh Mookerjee "with warmest admiration and esteem".

One must pause and absorb the significance of the various questions Raman raised in his essay. Right from 1905, Einstein had been advocating the idea that radiation energy was quantised. (Much later in 1925, these quanta were named *photons* by the American chemist Gilbert Lewis.) Based on the idea of the light quantum, Einstein had successfully explained the phenomenon of photoelectricity, and Millikan, famous for his work on the charge of the electron, had verified Einstein's photoelectric equation in all its aspects (see Chapter 3 of *Bose and His Statistics*). Nevertheless, few would accept the idea of the light quantum – such was the power of the classical theory of electrodynamics due to Maxwell. In fact in that very year when Raman wrote his essay (1922), Einstein received the coveted Nobel Prize for his work on photoelectricity but of the light quantum, not a single word in the Nobel citation! So it was somewhat bold of Raman to think in terms of a quantum theory of light scattering when leading lights like Bohr and Planck disapproved of even the idea of a light quantum.

4.5 Feeble fluorescence

Unaffected by what others thought, Raman now launched a vigorous programme of research to study "the possibility that the corpuscular nature of light might come into evidence in scattering". When Compton discovered the Compton effect (see Chapter 3 of *Bose and His Statistics*) in 1923 all doubts about the light quantum vanished overnight and it suddenly became meaningful to talk about the scattering of light quanta by atoms and molecules. Compton had found out what happens when X-ray quanta are so scattered. Raman now began to wonder what would happen if the frequency of the quantum was low enough to be in the visible part of the electromagnetic spectrum.

During the next few years, light scattering took precedence over all other research at the Association. Now that there were plenty of students around, they were all pressed into various investigations and experimental data poured out.

When light is scattered, two things connected with it could get affected – the frequency and the polarisation. According to classical electrodynamics, light scattering with a change of frequency was NOT possible. So light scattering in classical theory always meant scattering *without change of frequency*. Rayleigh scattering belongs to this category. Of course the polarisation could change, but that was a different matter.

For a moment, suppose there is a change of frequency of light consequent to scattering. Without bothering about how or why such a change occurs, let us merely ask: If such a frequency change does occur, how does one recognise it? The obvious solution to this problem is indicated in Fig. 4.1. This technique is straightforward

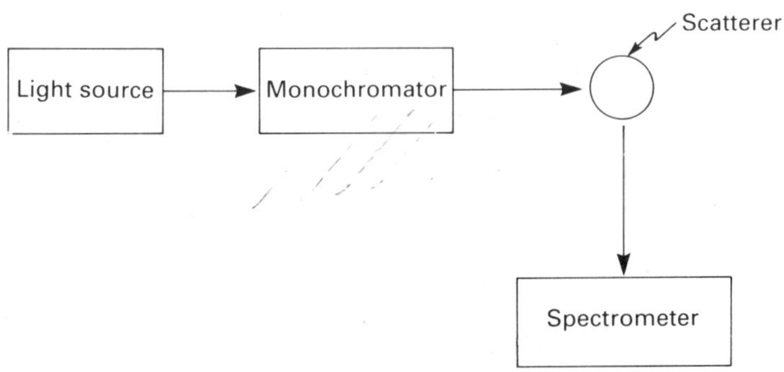

Fig.4.1 The textbook arrangement for observing light scattering accompanied by a change of frequency. Today, the light source is usually a laser. In the pre-laser era, it was usually a mercury-arc lamp.

and is the one used today. However, for this technique to work, the scattered light must be sufficiently strong, which in turn means that the incident light must be strong.

This raises the question: What does one use for a light source? Today there are various types of powerful light sources available which emit radiation at different frequencies but in those days these types of sources did not exist. The few that were available cost money, and money was not exactly plentiful in the Association. However, Raman was not put out by all this. He simply decided to use sunlight which was available in plenty and came free!

Figure 4.2 shows a simple scheme for piping sunlight into the

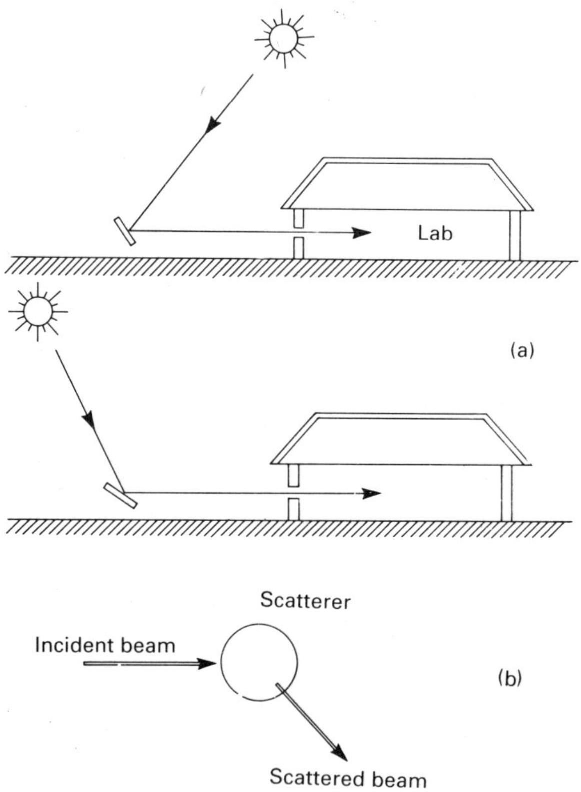

Fig.4.2 Concept of the heliostat. It essentially consists of a mirror which reflects sunlight in the same direction. In the last century when artificial and powerful light sources did not exist, the heliostat was the only means available. Here **(a)** shows how the mirror is suitably tilted to keep the reflected beam along the same direction. **(b)** shows how the beam could be used in the lab.

laboratory. One essentially uses a mirror to reflect the sunlight. But there is a problem because the Sun keeps moving in the sky. However, the motion is quite regular and therefore if one systematically adjusts the mirror, one can make sure that the light is always piped in, even though the Sun is moving. One does not necessarily have to stand in the hot sun and adjust the mirror by hand every once in a few minutes. Instead one could rig up a simple mirror-tilting mechanism driven by clockwork which does the tilting automatically. Such a mechanically-operated mirror (called heliostat) was available to Raman and he used it regularly.

The use of sunlight solves the source problem in a manner of speaking, but it brings up another which is that the scheme of Fig.4.1 cannot be used. Sunlight is not powerful enough. How then does one find out if the colour of light has changed on scattering? Raman used his ingenuity and invented a method called the method of *complementary filters*. The idea is shown in Fig. 4.3.

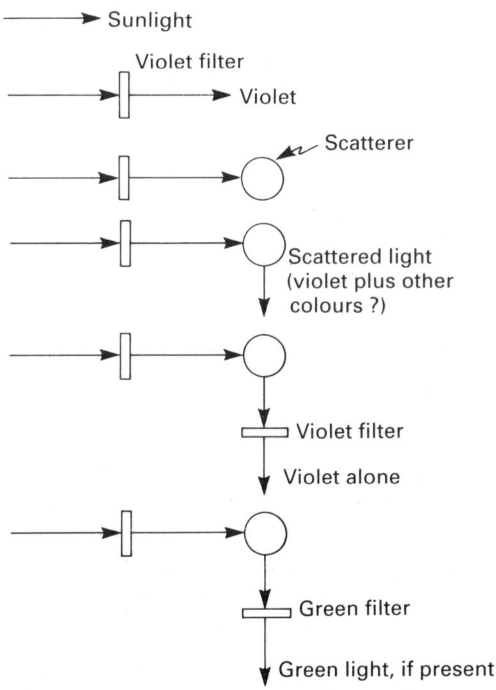

Fig.4.3 Arrangement of complementary filters devised by Raman. Observations are made visually. The advantage is that a number of scatterers can be examined rapidly.

Suppose one places say a violet filter in the path of the sunlight. This would produce a violet beam, i.e., out of all the colours present in the white light, violet alone is allowed to pass. This beam of violet light is now allowed to fall on a bulb containing some liquid. The liquid scatters light in all directions but say we observe the light scattered at 90°. There are now two possibilities; either the scattered light also is violet or it is not. Actually, there is a third possibility which is that part of the scattered light is violet and part of it is not. Suppose we put a violet filter in the path of the scattered light. Then only the part that is violet will pass through the filter. If, on the other hand, we place a green filter, the violet part of the scattered light would be cut out and only the green part (if any) would be transmitted.

If we want to quickly find out if there has been a change of frequency, the strategy is clear. (i) Pass sunlight through say a violet filter and allow it to fall on a bulb containing some liquid. (ii) View the bulb at 90°; this corresponds to looking at light scattered at right angles. (iii) Introduce a green filter and view. If green light is seen, then part of the incident violet light has been scattered as green light, i.e., there is change of frequency. If no green light is seen, there is no change of frequency.

You might ask: "What if the scattered light is yellow and not green?" No problem; just use a yellow filter instead of a green one to check it out. Raman used violet and green filters and since the two colours are complementary, the method was called the method of complementary filters.

So what did they find with this method at the Association? Almost right from the beginning, it was noticed that there was some evidence of light being scattered with a change of frequency, irrespective of the substance that did the scattering. True, this type of scattering was quite feeble but it could be seen nevertheless. Notice that observations were made visually. How does one see light that is feeble and be sure about it? Obviously, one had to have good eyesight – incidentally, Raman appears to have been particularly blessed in this respect. He had the uncanny gift of being able to see very feeble optical phenomenon quite clearly, as is evident from the various descriptions he has recorded at various points of time. Getting back to the observation business, at Calcutta they had an interesting method. Let us hear Raman describe it:

> As it is very essential to prevent the distraction of the eye by external light and to keep the eye at its maximum sensitiveness to feeble illumination like that due to scattering, the observer located himself

in a light-tight wooden cage 4 ft square [$\sim 1.2 \times 1.2$ metres] and 8 ft [2.5 metres] high, facetiously known as the 'Black Hole of Calcutta'. [For information about this infamous Black Hole, see Box 4.2]

Box 4.2 The "Black Hole of Calcutta" incident is supposed to have occurred in 1756 during the clash between the forces of the East Indian Company and Siraj-ud-daulah, the Nawab of Bengal. Siraj-ud-daulah, stormed Fort William and the garrison there under Holwell surrendered. British historians claim that the Nawab locked up a large number of British soldiers in a small room following which many died due to suffocation. However, Indian historians have hotly contested this claim and say the charge by the British is a trumped up one.

Incidentally, Lord Rutherford and his students also relied on visual observations in their epoch-making experiments on alpha-particle scattering which led to the discovery of the atomic nucleus. Believe it or not, the eye, having adjusted to darkness, is a very good photon detector – maybe not as good as the photomultipliers available today, but a very good one nevertheless.

Let us get back to the findings of Raman and his students. K.R. Ramanathan was the first to notice that colour change occured on scattering. He made this observation way back in 1923. How does one explain the observation was the next question. At the time, the only type of light scattering known was Rayleigh scattering and that did not allow for a change of frequency on scattering. Now before concluding that there was indeed a new type of scattering accompanied by a frequency change, one had first to make sure that there was no other explanation possible for the observed phenomenon.

What was the observed phenomenon? It was that when the liquid in the bulb was illuminated with light of one colour and was observed in a perpendicular direction, light of another colour was seen. Can one be sure that the light so seen is really *scattered* light and not light produced by some mechanism other than scattering? Is there any such mechanism? Yes there is, and it is called *fluorescence.*

Starting with K.R. Ramanathan in 1923, many researchers at the Association repeatedly saw a feeble intensity at a frequency different from that of the incident illumination. And they all interpreted it as being due to fluorescence. Normally, fluorescence is produced by impurities in the material. So the researchers tried to purify the material. However, no matter how much the purity was improved, the mysterious extra component (other than the Rayleigh scattered light) remained; clearly, it was not being caused by impurities. No one could explain it and it was simply described as "feeble fluorescence", although no one was sure that this component was indeed due to fluorescence.

4.6 What is fluorescence?

Many materials have the ability to absorb and re-emit light. Suppose a material is illuminated as in Fig.4.4(a). And let us say that this material can absorb and re-emit light. Typically the re-emission intensity would vary in time as in Fig. 4.4(b). Part of the re-emission goes on while the incident light is on but the interesting thing is that light emission continues even after the source is turned off. You might

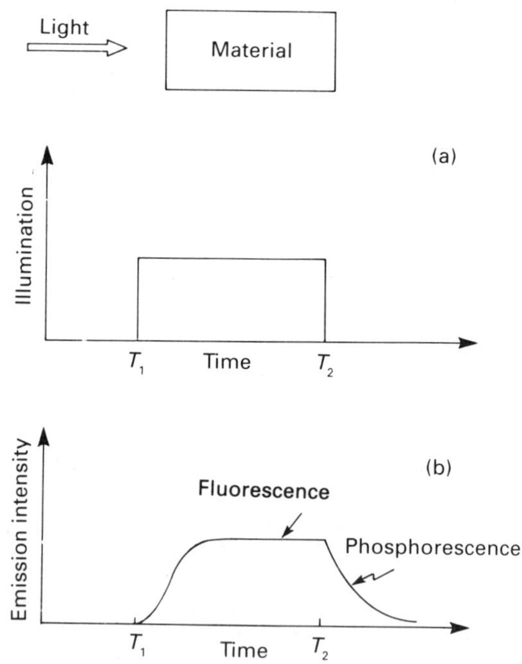

Fig.4.4 Time behaviour of emission of light from an illuminated material. (a) shows the time profile of the incident light. It is on for a certain time. (b) shows how the emitted light intensity varies with time. Observe that emission continues even after the incident beam is switched off.

have seen the screen of your TV glow for a few seconds after the set is switched off. The phenomenon I am talking about is something similar. The re-emission of radiation during illumination is called *fluorescence* and the emission after the incident light is off is called *phosphorescence*; the term *luminescence* is often used to cover both.

Figure 4.5 gives a schematic explanation of fluorescence in terms of an energy level diagram. Incident light is first absorbed by the system, say a molecule of the liquid that is being illuminated, and as a result the molecule goes to a state of higher energy. The molecule then loses some energy, perhaps by colliding with other molecules, and by this process it goes down to level m. The molecule then makes a transition to the ground state by emitting a quantum of

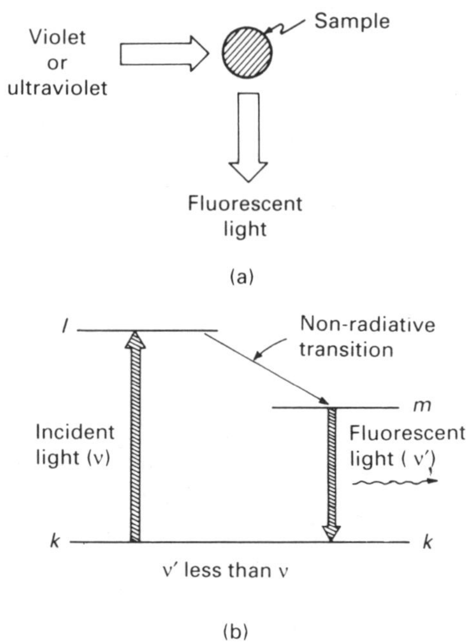

Fig.4.5 (a) Schematic of the fluorescence experiment. (b) Energy changes during the process. First there is absorption of a light quantum (of frequency ν) causing the atom to change its level from k to l. This is followed by a non-radiative transition to level m from where the atom/molecule returns to the ground state k emitting a photon of frequency $\nu'(\nu' < \nu)$.

light. It is evident from the diagram that the energy of the emitted quantum would be less than that of the absorbed quantum. In other words, the fluorescent light would have a lower frequency than that of the absorbed light.

Notice there are two steps in the process of fluorescence: First there is a clear absorption of the incident light and then there is re-emission (at a lower frequency). Fluorescence is thus quite distinct from scattering.

4.7 Optical analogue of the Compton effect

From the time when K.R. Ramanathan first spotted it, the mysterious phenomenon of "feeble fluorescence" was repeatedly seen by many researchers. In 1925, Krishnan saw it in as many as fifty liquids. No one knew what it was and no one was truly convinced that the extra component of light was produced by fluorescence. However, it was simply called by that name for want of a better word.

Raman, who had been thinking hard about this mysterious "feeble fluorescence", now began to seriously consider whether this was connected with the quantum nature of light. Compton's experiments had convincingly established that photons existed, exactly as Einstein had predicted a long time ago. And Raman had already speculated in his essay of 1922 that the quantum aspect of light would be important in scattering. Now his thoughts began to sharpen and he knew what to look for. In the Compton effect, the photon changed its frequency on scattering. It is true that the Compton effect dealt with X-ray photons. Could something similar be happening with photons having a frequency in the visible region too?

Even as new ideas were taking shape in Raman's mind, S. Venkateswaran, a researcher studying light scattering from glycerine reported that "the colour of sunlight scattered in a highly purified sample of glycerine was a brilliant green instead of the usual blue." Raman was now tremendously excited and brought K.S. Krishnan into the act. Just around this time, the news was received that Compton had been awarded the Nobel Prize. And it had an electrifying effect on Raman. Ramdas tells that story:

> On a December evening in 1927 when I was in Calcutta after an examination and an interview at Delhi, I visited the Association at Calcutta to pay my respects to Prof.Raman. His elder brother Sri Subramania Iyer was also with Prof.Raman when I entered his office. Soon after I went in, K.S. Krishnan rushed in and excitedly informed Professor of the announcement in the evening papers that

Prof. A.H. Compton had been awarded the Nobel Prize ... On hearing this news, Prof. Raman beamed with delight and burst out in his characteristic fashion: "Excellent news ... very nice indeed. But look here Krishnan. If this is true of X-rays, it must be true of light too. I have always thought so. There must be an optical analogue to the Compton effect. We must pursue it and we are on the right line."

4.8 The great discovery

Events now began to move really fast. The best way of describing them is to give you extracts from the diary of K.S. Krishnan.

February 5, 1928

For the last three or four days, I have been devoting all my time to fluorescence. The subject promises to open out a wide field for research, since at present there is no theory of fluorescence which could explain even the outstanding facts...

Recently, Prof. has been working with Mr. Venkateswaran on the fluorescence exhibited by many aromatic compounds... the fluorescence of some of the liquids is found to be strongly *polarised*...

February 7, Tuesday

...discovered that all pure liquids show a fairly intense fluorescence in the visible region and what is more interesting, all of them are strongly polarised...

When I told Prof. about the results, he would not believe that *all liquids* can show *polarised* fluorescence... He was very much excited and repeated several times that it was an amazing result... He wondered how we missed discovering all that five years ago...

After meals at night, Venkateswaran and myself were chatting together in our room when Prof. suddenly came to the house (about 9 p.m.) and called for me. When we went down, we found he was very much excited and had come to tell me that what we had observed that morning must be the Kramers–Heisenberg effect we had been looking for all these days. We therefore agreed to call the effect *modified scattering...*

February 9, Thursday

Set up this morning the long telescope and made preliminary arrangements for observing the effects with vapours. Before the arrangements were completed, Prof. left for the college for his lecture...

When Prof. came from the college at about three, I announced to him the result and there was still enough sunlight for him to see for himself. He ran about the place, shouting all the time that it was a first rate discovery, that he was feeling miserable during the lecture because he had to leave the experiment...

February 16, Thursday

Studied today pentane vapour at high temperature and it showed a conspicuous polarisation in the modified scattering. We sent a note today to *Nature* on the subject under the title "A new type of secondary radiation".

February 17, Friday

Prof. confirmed the polarisation of fluorescence in pentane vapour. I am having some trouble with the left eye. Prof. has promised to make all the observations himself for some time to come.

February 27, Monday

Religious ceremony in the house. Did not go to the Association.

February 28, Tuesday

Went to the Association only in the afternoon. Prof. was there and we proceeded to examine the influence of the incident wavelength on the phenomenon. Used the usual blue-violet filter coupled with uranium glass, the range of wavelengths transmitted by the combination being much narrower than that transmitted by the blue-violet filter alone. On examining the track with a direct vision spectroscope, we found to our great surprise that the modified scattering was separated from the scattering corresponding to the incident light by a dark region [see also Fig. 4.6].

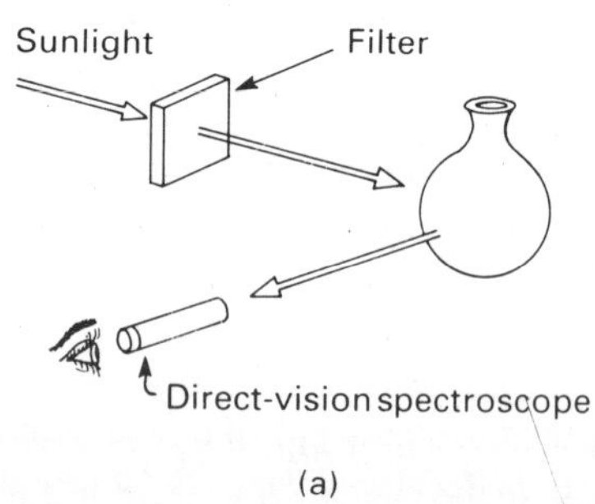

Fig.4.6(a) Arrangement used by Raman and Krishnan.

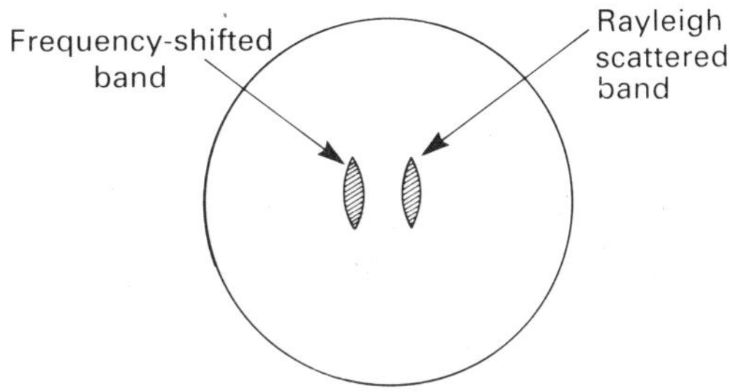

Fig.4.6(b) Schematic of the bands seen by Raman and Krishnan in a direct vision spectroscope.

4.9 Some comments

I hope you have closely followed the exciting events that culminated in the big discovery. For years, the so-called "feeble fluorescence" had been around. People repeatedly saw it, recorded its presence, maybe wondered a bit and then left it alone. Towards the end of 1927, it was noticed that this mysterious radiation was *polarised.* Fluorescence was NOT supposed to be polarised. That was what got Raman excited in the first place. So, not only was the mysterious radiation shifted in frequency from the incident radiation, but it was also polarised.

One can now make a table as follows (see also Fig.4.7):

Phenomenon	Polarisation	Freq. change
Rayleigh scattering	Yes	No
Fluorescence	No	Yes
"Feeble fluorescence"	Yes	Yes

From this table it is clear that the so-called "feeble fluorescence" was quite different from the normal Rayleigh scattering as well as from the usual fluorescence. What then was it? This is where Raman's remark about the Kramers–Heisenberg effect is significant

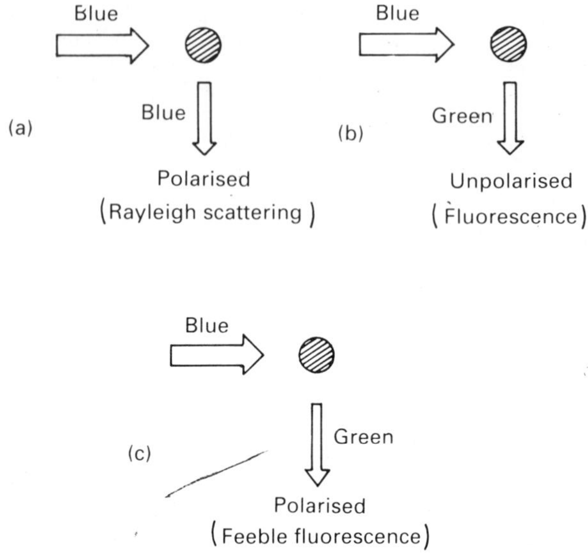

Fig.4.7 Schematic illustration of the differences between (a) Rayleigh scattering, (b) usual fluorescence and (c) the mysterious "feeble fluorescence". Eventually, the latter was identified as what we today call Raman scattering.

(see diary entry for February 7). I have discussed this work of Kramers and Heisenberg in the companion book *The Quantum Revolution: Part* I, and I shall consider it again in the next chapter. Basically it refers to a new type of scattering permitted only in quantum mechanics. Raman sensed that what his students popularly referred to as "feeble fluorescence" was not fluorescence at all but a new type of scattering; that is why he decided to call it *modified scattering*. Note that Raman recognised this radiation as being due to scattering. Today, we refer to it as *Raman scattering*. It is a type of scattering in which radiation of frequency ν, say, changes to a different frequency ν' say, after scattering. The experiment performed on February 28 clearly brought this out. The incident radiation was made nearly monochromatic by using the combination of the usual blue-violet filter and a piece of uranium glass. The scattered light was examined with a pocket spectroscope. And what Raman and Krishnan saw was a strong band corresponding to the incident light and another band, somewhat feeble, but clearly visible nevertheless, which was well separated from the band corresponding to that associated with the incident radiation – recall Krishnan's reference to the dark band separating the two.

The *Raman effect* had finally been discovered and on the following day, Raman announced his discovery at a press conference. In commemoration of this event and also to inspire us in our scientific enquiry, we now celebrate February 28 as the National Science Day. Two weeks later, to be precise, on March 16, Raman gave a detailed lecture on his discovery at the South India Science Association at Bangalore. And in that lecture he showed for the first time, slides of the frequency shift produced by Raman scattering. These were obtained using the arrangement in Fig. 4.8.

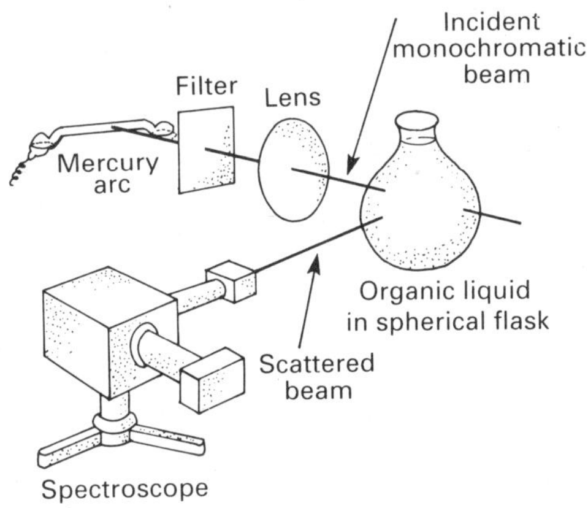

Fig.4.8 Arrangement used by Raman to photograph Raman shifts in several liquids.

On his way to Bangalore, Raman stopped at Madras at his brother's place and about this visit, Chandrasekhar recalls:

> I have a vivid recollection of a day in early March of 1928, when Professor Raman visited our home... I remember well his showing slides of the first Raman spectra ever taken and the state of euphoria he was in. On that occasion someone drew his attention to the discovery of the Compton effect some few years earlier, and Raman responded with, "Ah, but my effect will play a very great role for chemistry and molecular structure!" That statement was indeed prophetic.

4.10 The honours

It was clear from the very beginning that Raman's discovery was a very important one. From the point of view of quantum physics, it not only gave further proof of the quantum nature of light (which of course pleased Einstein very much) but also confirmed a new type of scattering foreseen by Kramers and Heisenberg. And on the applications side, it literally opened the flood gates; the flood is yet to abate!

Raman's discovery was hailed by one and all. Rutherford made a reference to it in his Presidential address to the Royal Society in 1929. The British Government conferred a knighthood, various societies honoured Raman with medals, universities conferred honourary doctorates and so forth. The climax of course was the announcement of the Nobel Prize.

Raman was pretty sure he would bag the Prize. In fact he expected the award in 1928 itself and was disappointed when it went to Richardson. In 1929 also he did not receive the award; that year the winner was de Broglie. Quite confident that he would definitely make it in 1930, he booked two steamer tickets to Europe for himself and his wife as early as July, even though the awards were to be announced only in November. Raman's point was that by then it might be too late to get reservations, and he would then have to miss the award function which usually took place in December. I don't know whether you would call it confidence or ego – anyway, there it was!

Booking tickets is fine but then one must actually get the Prize. Raman naturally spent many anxious months awaiting the announcement. Bhagavantam told me how he and Raman would go out for a walk every morning, how Raman would buy the newspaper and eagerly scan for an announcement and how he (Raman) would toss the paper away if he did not find any news from Stockholm. Eventually when the announcement was made, Bhagavantam received the phone call (from a local news agency) and had the pleasure of conveying it to Raman.

Lady Raman has given a nice account of the award ceremony.

> The award of the Nobel Prizes took place on the 10th [December] between 4 and 7 p.m. If I shut my eyes, I can still see as in a dream the great concert hall of Stockholm, decorated with flowers and flags, filled with more than 4000 people, the King and the Queen of Sweden and the Royal family occupying the first seats. The Nobel Laureates then entered the hall... The Secretary of the Academy of Sciences read a report and gave a brief account of Nobel's life. Dr.Pleijel, Professor of Electro-Technics in the University of Stockholm, then

> rose and spoke for twenty minutes on my husband's investigations on the scattering of light and the new effect that had been discovered by him, and addressing him said: Sir Venkata Raman, the Royal Academy of Sciences has awarded you the Nobel Prize in Physics for your eminent researches on the diffusion of light and for the discovery of the effect that now bears your name. The Raman effect has opened new routes and has given most important results. I now ask you to receive the prize from the hands of His Majesty.

Present in the audience was also the Charge d'Affaires of the United States. Since two Americans received the Nobel award during the same function (Carl Landstein for Medicine and Sinclair Lewis for Literature) he wrote a report for his boss in the U.S. State Department. In it he said:

> Of all the prize winners, the day was easily carried, however, by Sir Venkata Raman the Indian Prize winner, who, upon returning to his seat on the platform after receiving the prize from the hands of the King, was visibly moved by his emotion and sat with tears streaming down his face...
>
> At the banquet that evening, Sir Venkata Raman's speech was a masterpiece of eloquence which called forth tremendous applause from a banquet-weary gathering not noted for its responsiveness. Less appreciative was perhaps the British Ambassador, who sat one place removed from me, who was forced to listen with equanimity to Sir Venkata Raman's reference, brief though it was, in passing only, to the congratulatory telegram which he received "from his dearest friend now in jail".

All this might appear strange to you and so let me add a few words of explanation. Firstly, on the platform, you would see displayed, the flags of the countries to which the winners belong. In 1930, India was still a British colony. We had no flag of our own and our flag was nothing but the British flag – the Union Jack. "The dearest friend" Raman referred to was none other than Mahatma Gandhi, who at that time had been put in prison by the British.

Raman himself has described those moments, especially his breaking down. He recalls:

> When I sat down in that hall and I saw the sea of western faces surrounding me, and I, the only Indian, in my turban and closed coat, it dawned on me that I was really representing my people and

my country. I felt truly humble when I received the Prize from King Gustav; it was a moment of great emotion but I could restrain myself. Then I turned around and saw the Union Jack under which I had been sitting and it was then I realized that my poor country, India, did not even have a flag of her own – and it was this that triggered off my complete breakdown.

4.11 Who got there first?

Raman was not the only one interested in light scattering at that time. There were other active groups, especially in France and in Russia. I must here make a brief reference to work done by the Russian physicists Landsberg and Mandel'shtam. The latter was the senior member of the pair, and he had been interested in various aspects of light scattering from 1913 onwards. Serious experimental work was commenced with Landsberg starting in 1925. The Russians chose quartz as the system for study. They decided on quartz because they thought that large samples of good quartz would be easy to procure (large samples were preferred as that would increase the intensity of the scattered light). Landsberg used to go shopping for quartz but unfortunately he was repeatedly cheated by clever vendors who palmed off inferior quartz as being of high quality. If the quartz contains defects or impurities, it leads to background scattering which masks the effect sought for – i.e., scattering with a frequency shift. Poor Landsberg lost a lot of money in the process. However, by early 1928 a good specimen of quartz was finally procured.

Landsberg and Mandel'shtam were looking for a type of scattering called Brillouin scattering (which is explained in the next chapter). Their experimental arrangement is shown in Fig. 4.9; compare this

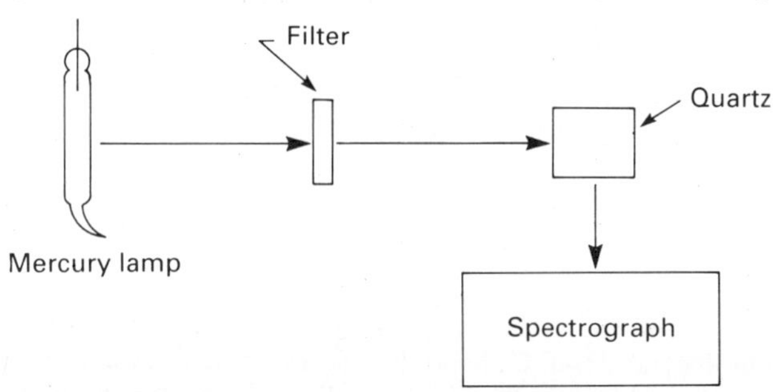

Fig.4.9 Experimental arrangement used by Landsberg and Mandel'shtam. Compare with Figs.4.1 and 4.8

with Figs. 4.4 and 4.5 – no sunlight, complementary filters, etc.; a straightforward arrangement in fact.

Once he tasted success, Raman too employed a similar arrangement as I described in Fig. 4.8; this is how he produced the spectra which he showed during his lecture at Bangalore. In the case of liquids, the scattering is quite strong and one can record a spectrum relatively easily. In quartz the scattering is comparatively weak, and Landsberg and Mandel'shtam had to give an exposure of over 15 hours to record a spectrum. They first noted the appearance of new lines on February 21, 1928. The Russians did not publish their results but reported it during a conference in Moscow in April. Unfortunately, few believed in their result and they felt discouraged. Meanwhile, Raman had announced his discovery and the Russians realised that they too had made the same discovery, without having published their result. They then tried to make up for lost time by sending a paper, which they did in May. By this time, sixteen papers on the Raman effect had already appeared, including several by Raman himself.

When the Nobel Prize was awarded to Raman, the Russians were sore. Many people in Russia refused to call the effect by the name it was generally known, i.e., the Raman effect, even though the rest of the world did so. But time is always a great healer, and today that disappointment is a distant memory. Raman hunted for and first caught the effect *visually*, using the method of complementary filters. In fact when he gave his famous lecture in Bangalore on March 16 (see Section 4.9), he had the visual arrangement set up and made the members of the audience experiment with filters and see the colour change. It was only when he had seen this effect appearing in a large number of liquids that he went in for a conventional data recording experiment.

What is the physics underlying the Raman effect? For that, turn to the next chapter.

5 *Physics Of The Raman Effect*

In this chapter, we shall consider the physics of the Raman effect. As you might have gathered from the last chapter, the Raman effect is quantum mechanical in origin. Before we come to that, we must first understand the classical picture of light scattering.

We start by discussing the conditions under which light scattering may be expected. Perhaps you would be surprised to learn that a perfectly homogeneous medium will not scatter light. Now what do I mean by a perfectly homogeneous medium? Essentially a material whose density is the same everywhere. Actually, I should be a bit more careful and say: the same *optical* properties everywhere. Shortly, you will understand the meaning of this fineprint.

Consider a block of homogeneous matter as in Fig. 5.1. This could be a block of clear glass, without any impurities, bubbles, etc. Let us say we shine a ray of monochromatic light into this medium as shown. We now consider an elementary cube P whose dimensions are smaller than the wavelength λ of the light. Assume that every part of this elementary cube scatters light. Since the cube size is small, the different portions of P will scatter light in phase

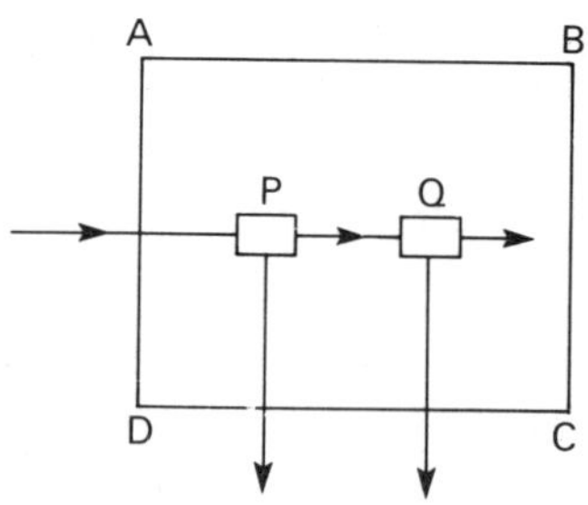

Fig.5.1 Figure to illustrate the argument that a homogeneous medium cannot scatter light. Explanation in text.

with each other. Thus at a faraway point at a distance x from P, the amplitude of the scattered wave can be written as

$$A\cos\{\omega(t-x/c)+\theta_x\}. \tag{5.1}$$

Here ω is the angular frequency ($= 2\pi\nu$), t the time of observation, c the velocity of light and θ_x the phase angle. In the same manner, another elementary block Q would produce at x an amplitude

$$A\cos\{\omega(t - x/c) + \theta_x - \phi\}. \tag{5.2}$$

Here ϕ is the phase difference between the light coming from P and from Q. If the block ABCD is large enough, we can always choose P and Q such that $(\phi - \theta_x) = \pi/2$. The waves from P and Q would then cancel each other. In this way, all the secondary or scattered waves coming from different regions of the medium could be paired off, leaving no aggregate scattering in the direction x. The same argument could be used for every other direction. In short, the block ABCD is unable to scatter light and this is so because all the elementary regions are strictly alike in their optical properties. If they were not, the amplitudes of all scattered waves would not be A. In this case one would see some scattering which would also be different in different directions.

You might ask: "What about elementary regions close to the surface? Can they all be paired off? Can we not expect some scattering from them?" Indeed there would be some surface scattering but if the size of the block is sufficiently large, the surface to volume ratio would be very small and one could perhaps neglect the surface effects. Anyway, the point is not about the surface but about what happens inside. And as far as the inside is concerned, there is no net scattering; that is the real point.

Suppose we replace the glass with a bulb containing pure liquid, say water. What happens now? This time there would be some scattering. How come? The water is absolutely pure and so, will not the arguments made earlier for glass apply here too? Not quite. Why? Because the local density can fluctuate—see Fig. 5.2. Thanks to these fluctuations, the optical properties of the water in the bulb would not be the same everywhere and as a result there would be some scattering. Of course, the scattering would be quite weak, as Raman first noted. It might be useful to quote from his paper to the Royal Society (recall last chapter). He talks there of how he started his experiments with "ordinary town-supply water", i.e., tap water. He says:

> The track [of the incident beam] was practically white and showed innumerable motes floating about in the water. Repeated filteration

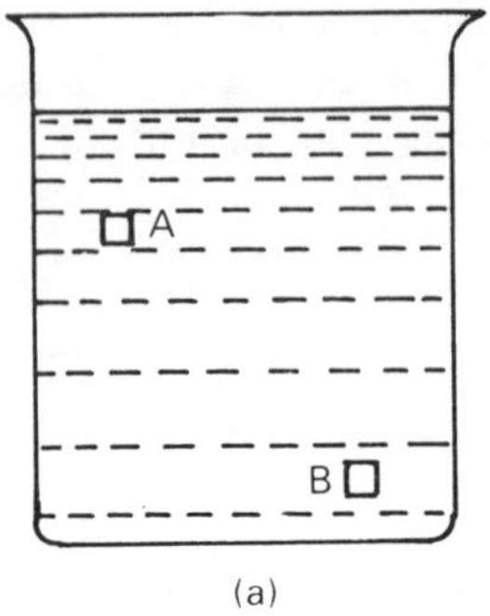

Fig.5.2 Consider a beaker of water as shown. Focus on a small volume around a point like A. Since molecules are in constant motion, the number of water molecules in this elementary volume would fluctuate randomly in time. There would therefore be corresponding density fluctuations. The fluctuation would be similar at all points, assuming that the temperature is uniform. These density fluctuations produce corresponding fluctuations in refractive index, and it is the latter which produces the optical inhomogeneity needed for light scattering.

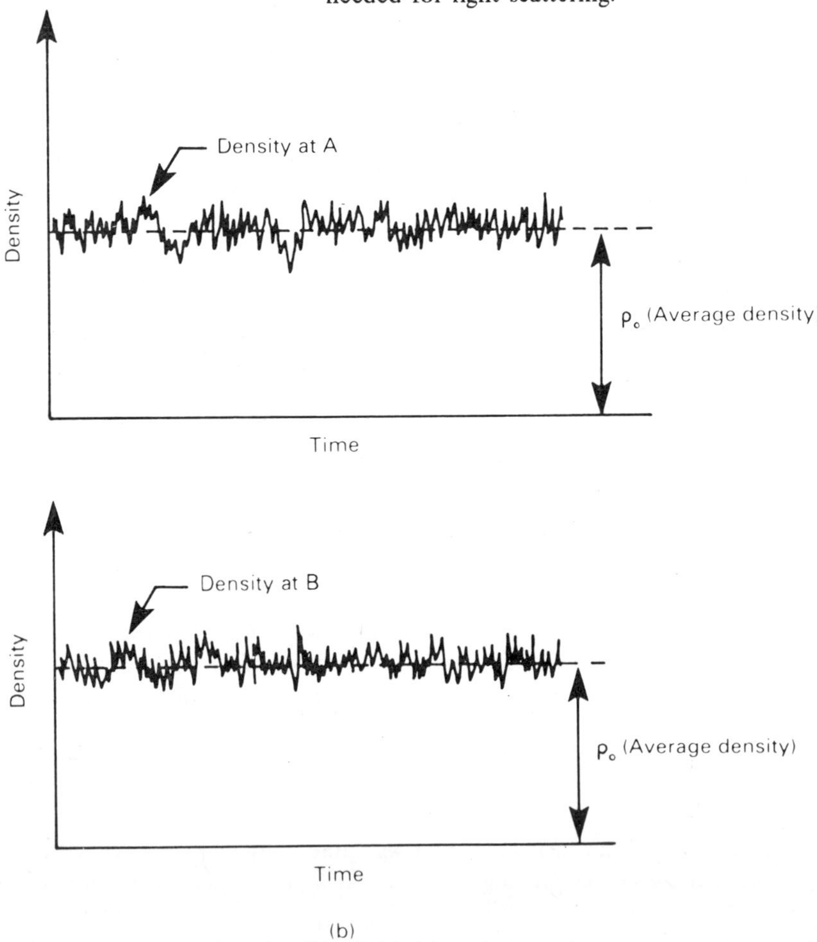

> through several thicknesses of Swedish filter paper made an improvement... A somewhat casual attempt was then made to clear the water by adding alkali and alum thus throwing out a gelatinous precipitate of hydroxide. This made a further improvement... The next attempt was made with ordinary distilled water... This gave immediately a much smaller intensity of scattering than the tap-water had done after several attempts at filteration.

Normally, fluids do not scatter light strongly. But near a second-order phase transition, they do. I have discussed this in the companion volume *The Many Phases of Matter* and shall therefore not repeat myself here. But there are a few other things that I would like to mention.

The strong scattering of light near a phase transition is referred to as *critical opalescence*. Einstein and Smoluchowski were the first to give an explanation for it, and they did it independently of each other. Basically, what they said was the following: Near a phase change, the density fluctuates strongly within the medium. As a result, the refractive index also fluctuates from point to point. And the net result is that there is strong light scattering. The intensity of the scattered light is conveniently described in terms of a quantity K called the scattering coefficient. For transverse scattering (i.e., at 90^0), it is given by the formula

$$K = \frac{\pi^2}{18}\frac{RT\beta}{N_A\lambda^4}(n^2-1)^2(n^2+2)^2. \tag{5.3}$$

Here λ is the wavelength of the incident light, R the gas constant, n the (average) refractive index, T the absolute temperature, N_A the Avogadro number and β the compressibility. There are two things I would like you to notice. Firstly, the scattering coefficient is inversely proportional to the fourth power of the wavelength (in other words, proportional to the fourth power of the frequency ν since $\nu\lambda = c$). Secondly, the scattering is proportional to the compressibility. Near a phase change, β increases enormously and this is what is responsible for critical opalescence.

Einstein and Smoluchowski derived formula (5.3) for explaining critical opalescence, starting from density fluctuations because they hold the key. The density does fluctuate rather strongly near the critical temperature T_c but that does not mean it does not fluctuate at other temperatures. In fact it does, which means that light must be scattered by a fluid even far from the critical temperature. This precisely was Raman's hunch during the journey across the Mediterranean, and he verified it with experiments performed soon after his return to Calcutta. Raman not only proved that even ordinary water scatters light but also that the result could be broadly understood in terms of the Einstein–Smoluchowski formula itself. As we have seen, according to that formula the scattering is proportional to the fourth power of the frequency. Thus when sunlight falls on the water in the sea, it gets scattered. But of the incident white light, it is the blue which is scattered most because of the $(1/\lambda^4)$ or the (ν^4) factor.

It is interesting to ask what form formula (5.3) would take in the case of a perfect gas. Here $n \sim 1$, and $\beta = (1/p)$ where p is the pressure of the gas. One then gets

$$K = \frac{\pi^2}{\lambda^4} \frac{RT}{N_A p} (n^2 - 1)^2 \ . \tag{5.4}$$

This formula was first derived by Rayleigh, though by a completely different method. Rayleigh used this formula to explain the blue of the sky – he too leaned on the $(1/\lambda^4)$ factor to argue why the sky is blue. Raman was aware of this explanation and though Rayleigh's formula was for a gas, he knew about its linkage to the Einstein–Smoluchowski formula. That is why he refused to accept Rayleigh's observation that the sea is blue because it is merely reflecting the sky. Raman's preliminary observations made while travelling on the *S.S. Narkunda* convinced him that Rayleigh was wrong as far as the colour of the sea was concerned. He just couldn't wait to reach Calcutta and start on detailed experiments. And the results he obtained confirmed his guess.

In passing, I should mention that in solids too there is a movement of atoms and consequently some density fluctuations. However, it is much less than in a liquid. Therefore, solids like glass and ice scatter light to a much less extent than a liquid. To recapitulate, Raman knew of Rayleigh's scattering formula and how it helps to explain the blue of the sky. He also knew of the Einstein–Smoluchowski formula. Though this formula was derived to explain critical opalescence, Raman guessed it could be used far from T_c also. He further knew (5.3) and (5.4) were related. All this made him guess (5.3) could be used to explain the blue of the sea.

Raman did not stop with explaining why the sea is blue. He now became interested in the *mechanism* of light scattering. It is alright to say that light is scattered because the medium is optically inhomogeneous, i.e., the refractive index varies from point to point.

How is the refractive index dependent on the properties of the atoms and molecules making up the medium and how do these properties finally determine the scattering? This was the question Raman began to think deeply about and the essay *The Molecular Diffraction of Light* published in 1922 contained his speculations on the matter.

Now when a beam of light falls on, say, a liquid contained in a bulb, the light actually interacts with the atoms and molecules; and it is from these interactions with the individual atoms and molecules that one gets a transmitted beam, a refracted beam or a scattered beam. So it is essential to understand how light interacts with an atom.

There is a simple way of understanding this in classical physics. Here one uses the Lorentz model for the atom, i.e., one essentially regards the atom as in Fig. 5.3. Let $E_0 \cos \omega t$ denote the electric field

Fig.5.3 Lorentz visualised the atom as a harmonic oscillator, the oscillations being executed by the electron in the atom.

associated with the incident light. Due to this, a dipole moment is induced in the atom. The induced polarisation is given by $P = \alpha E_0 \cos \omega t$ where α is the polarisability of the atom. Observe that P oscillates in time. Thanks to this, the atom now radiates electromagnetic energy which is interpreted as the scattered radiation. Thus scattering is visualised as a two-step process: first a dipole moment is induced and next the induced dipole radiates.

The quantity α is given by

$$\alpha = \frac{e^2}{4\pi^2 m_e} \cdot \frac{1}{\nu_0^2}. \tag{5.5}$$

Here e is the electron charge, m_e its mass and ν_0 a characteristic frequency associated with the atom (see also the companion volume *The Quantum Revolution*: *Part* I for a further discussion of the polarisability). In a gas, the different atoms scatter independently of each other and so one might safely add the *intensities* contributed by the different atoms. If this is done, one arrives at the Rayleigh scattering formula. Incidentally, α is related to the refractive index n by $n^2 = (1+4\pi\alpha N)$ where N is the number of atoms per unit volume.

In the case of a liquid, the scattering contributions from the different atoms cannot quite be added like this because neighbouring atoms tend to form clusters (even if it be only for brief periods of time). They therefore scatter *cooperatively* or *coherently* and this effect must be taken into account. Essentially, the *amplitudes* from the neighbouring atoms must be added before squaring to obtain the intensity. Describing such correlations is not easy because one does not know much about how neighbouring atoms are arranged in a liquid. But Einstein and Smoluchowski circumvented these problems with some clever but gross descriptions, and it is this formula that Raman leaned upon while explaining his results on water.

The Einstein–Smoluchowski formula slurs over several things. From 1922 to 1927, Raman and his students therefore improved it in many ways (for example, by adapting to the case where the liquid contained nonspherical molecules). But notwithstanding these changes, all of which merely dealt with modifications to the scattered intensity, one important fact underlying this formula continued to remain the same, namely that it predicted scattering at the *same* frequency as that of the incident light.

As you know, from 1923 onwards, students of Raman were encountering another type of scattering which they dubbed "feeble fluorescence". Raman did not quite agree with this terminology, and as I described to you in the earlier chapter, Raman discarded the fluorescence theory when he found that this extra component was strongly polarised. What then was it?

This is precisely where Raman's remark to Krishnan (see Krishnan's diary entry for February 7) becomes significant. In 1925, Kramers and Heisenberg decided to take a close look at the scattering of light by atoms. This was part of an ongoing exercise to understand the intensity of spectral lines (see the companion volume *The Quantum Revolution: Part* I). They discovered that apart from Rayleigh scattering, there was the possibility of a new type of scattering. In their paper they wrote:

> When an atom is exposed to external monochromatic radiation of frequency ν, it not only emits secondary monochromatic spherical waves of frequency which are coherent with the incident radiation [we would call it Rayleigh scattering] but the correspondence principle also demands, in general, that spherical waves of other frequencies are emitted as well. These frequencies are all of the form $|\nu + \nu^*|$ where $h\nu^*$ denotes the energy difference of the atom in the state under consideration and some other state. The incoherent scattered

radiation corresponds in part to certain processes which have been recently discussed by Smekal...

Several comments are necessary. Firstly, even before the paper of Kramers and Heisenberg got published, Smekal had said that when radiation of frequency ν is scattered by an atom, the scattered radiation would not only contain a component of frequency ν but also other frequencies. What are they? Fig. 5.4 explains.

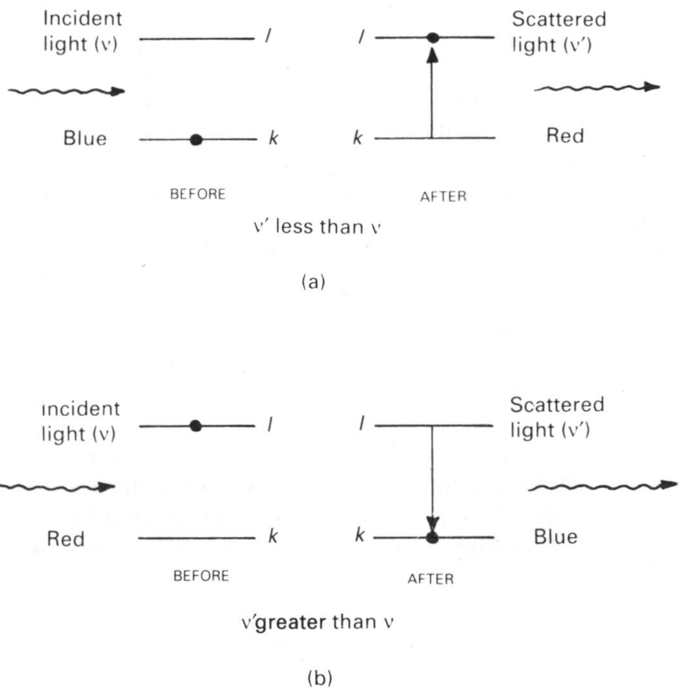

Fig.5.4 Schematic figure illustrating Smekal's view of how an atom could scatter a photon and cause a change in its frequency. Smekal considered only the possibility (a). Further explanations in text.

Basically, the incident photon loses some of its energy in exciting one of the atomic transitions, and emerges with a lower frequency. Smekal's analysis is easy to understand. Let M be the mass of the atom and v its velocity. After the collision with an incoming photon of frequency ν, let the velocity change to v'. Also, let the collision change the state of the atom from one having energy E_k to one having energy E_l. If ν' denotes the frequency of the photon after scattering, then from the principle of conservation of energy one has

$$(1/2)Mv^2 + E_k + h\nu = (1/2)Mv'^2 + E_l + h\nu'. \tag{5.6}$$

Smekal showed that the change of velocity was negligible so that one could write

$$E_k + h\nu = E_l + h\nu' \tag{5.7}$$

or

$$\nu' = \nu - (E_l - E_k)/h. \tag{5.8}$$

Thus the scattered light has a lower frequency than that of the incident light.

The analysis of Kramers and Heisenberg was much more detailed. Though their paper was published before the formal birth of quantum mechanics, much of their language was indeed what we would use today.

How does one understand the Raman effect in terms of modern quantum mechanics? I shall give you some idea, using the framework created by Dirac in 1928. In the trilogy on quantum mechanics, I refer to this as QED. I shall not go into details but if you recall, basically Dirac treated the problem of interaction of radiation with matter (in nonrelativistic QM) by considering the atom with its Hamiltonian H_a, the radiation with its Hamiltonian H_r and the Hamiltonian H_{int} which describes the interaction between the atom and the radiation. This interaction has the form $\widehat{\mathbf{p}} \cdot \widehat{\mathbf{A}}$ where $\widehat{\mathbf{p}}$ is the momentum operator and $\widehat{\mathbf{A}}$ that connected with the electromagnetic field ($\widehat{\mathbf{A}}$ is the vector potential, to be precise). There is actually another term $\widehat{\mathbf{A}}^2$ besides $\widehat{\mathbf{p}} \cdot \widehat{\mathbf{A}}$ but we ignore it here.

Raman scattering is described by second-order perturbation theory, i.e., in terms of the interaction $\widehat{\mathbf{p}} \cdot \widehat{\mathbf{A}}$ acting twice. Let us say that before the scattering, the atom is in the state $|l\rangle$ of energy E_l. I am here using the Dirac notation to describe the atomic state (this notation is explained in the trilogy on QM). The state of the radiation field is described similarly by stating how many photons there are of different frequencies. Here we shall focus only on the frequency ν_i of the incident radiation and the frequency ν_s of the scattered radiation. Let n_i and n_s be the number of photons corresponding to these two frequencies, in the initial state of the radiation field.

You might say that n_i should be equal to 1 and n_s equal to 0; I agree but I am writing these for the sake of generality. Thus the initial state of the radiation field can be written as $|n_i, n_s\rangle$; and the combined state of the atom plus the field would be $|l\ \rangle \cdot |n_i, n_s\rangle$.

What will the final state be? Let us say that as a result of the Raman scattering, the atom goes to the state $|k\rangle$ of energy E_k . The final state of the radiation field (as far as photons of frequencies ν_i and ν_s are concerned), would be $|n_{i-1}, n_{s+1}\rangle$; for the combined system the final state would then be $|k\rangle \cdot |n_{i-1}, n_{s+1}\rangle$.

The hop from the initial to the final state is made in two steps, taking help from the interaction term $\widehat{\mathbf{p}} \cdot \widehat{\mathbf{A}}$ twice – that is how second-order perturbation theory enters the picture. One can conveniently discuss this hop in terms of Feynman diagrams (for a description of these diagrams, see the companion volume *The Quantum Revolution: Part* II). The relevant diagrams are shown in Fig. 5.5. Observe there are two diagrams. You should not be rattled that in one of these, the scattered photon is emitted even before the incident photon has interacted. Actually, the time period between the wavy lines denotes what is referred to as the *intermediate*

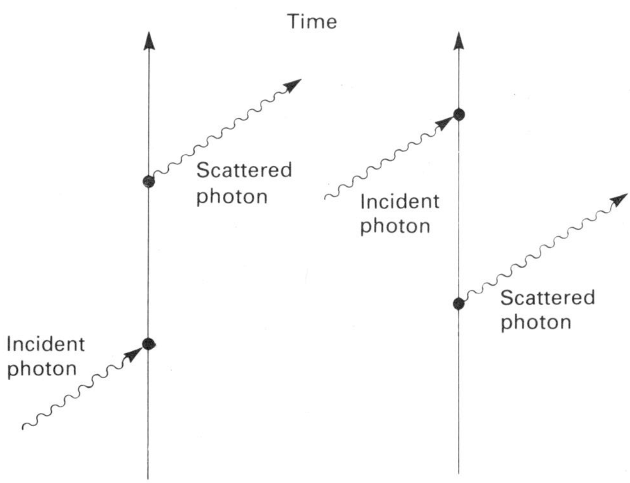

Fig.5.5 Feynman diagrams associated with the Raman effect. The region between the wavy lines represents the intermediate state.

state and we are not allowed to peek in there. Strictly speaking therefore, we are not entitled to talk of "before" and "after".

The atomic transitions associated with the scattering process are illustrated in Fig. 5.6. As a result of the perturbations, the atom first makes a transition from the initial state to some intermediate state $|m\rangle$ say; from there it makes a second transition to the final state. The intermediate state $|m\rangle$ could be anything and in fact we have to allow for all possibilities, i.e., sum over all intermediate states. One does not have to worry whether the incident photon energy $h\nu$ would be sufficient for such a transition to the intermediate state; the transition is *virtual* and therefore does *not* have to conserve energy (this is discussed in the volume on QED).

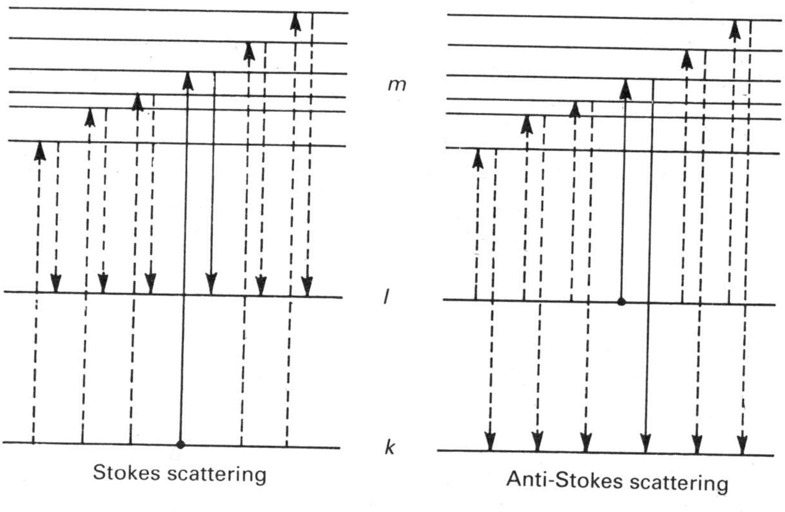

Fig.5.6 Various stages associated with Raman scattering, visualised in terms of the atomic energy levels. (a) refers to Stokes scattering ($\nu' < \nu$) and (b) to anti-Stokes scattering ($\nu' > \nu$). Observe that the transition from the initial to the final state occurs via a virtual intermediate state.

Notice there are two scenarios for the transitions as illustrated in Fig. 5.6. In one case the scattered photon has less energy than that of the incident photon. This is what Smekal had predicted, and the process is called *Stokes scattering*. In the other case, the photon emerges with more energy than it went in with by "borrowing" energy from the atom. Such a process is called *anti-Stokes scattering*. Kramers and Heisenberg were the first to predict such a possibility.

In May 1928, Raman and Krishnan observed this type of scattering for the first time and reported their result in *Nature*. However, they did not identify it correctly. That came a little later from Saha, Kothari and Toshniwal.

The important message underlying all that has been said so far is that Raman scattering refers to the scattering of photons in the visible region of the electromagnetic spectrum, accompanied by a change in the quantum state of the scatterer. In Fig. 5.6, the system we were considering was an atom and the transitions involved were electronic in nature. Therefore, this type of Raman scattering is called the *electronic* Raman effect. The study of electronic Raman effect in semiconductors acquired importance a few years ago. Incidentally, the transition

$$|l\rangle \cdot |n_i\rangle \rightarrow |l\rangle \cdot |n_i\rangle$$

is nothing but Rayleigh scattering; there is no change in the state of the scatterer and correspondingly also in the frequency of the photon.

In a liquid like water or carbondisulphide, there are molecules. Molecules can move (translate), rotate and also vibrate. Perhaps you are puzzled as to what is meant by vibration; this is illustrated in Fig. 5.7.

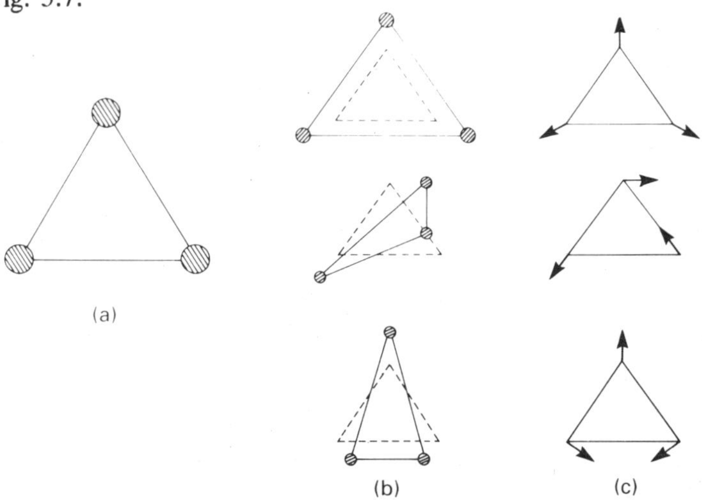

Fig.5.7 (a) shows the equilibrium configuration of the ozone molecule. The molecule has three normal modes of vibration. "Snapshots" of the molecule during these modes are shown in (b). The nature of the three modes are illustrated in (c). The lengths of the arrows are proportional to the amplitudes.

The rotational and the vibrational states of a molecule are quantised (see Box 5.1), and as result one can have *rotational Raman effect* and *vibrational Raman effect* respectively. What Raman's students were picking up as "feeble fluorescence" was nothing but light produced by vibrational Raman scattering. By the time he went to Bangalore to give his famous lecture, Raman was able to record nice spectra due to vibrational Raman scattering.

Box 5.1 A molecule is capable of three types of motion – translation, rotation and vibration; see figure (a). Associated with the rotation is a rotational angular momentum. Classically, this angular momentum can assume any value but in QM it is quantised. As a result, the rotational energy levels are discrete. Raman scattering accompanied by change of the rotational quantum state is called rotational Raman effect.

Consider now for a moment, Fig. 5.7(c). Select say one of the normal modes. Let its frequency be ν_0. The energy levels associated are given by $\epsilon_n = (n + 1/2)h\nu_0$. These describe the vibrational energy levels corresponding to the normal mode of frequency ν_0. Raman scattering accompanied by a change in the vibrational quantum state is called vibrational Raman effect.

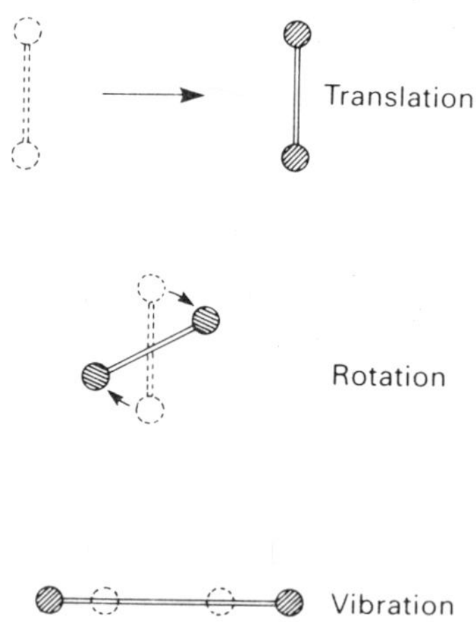

(a)

What about the translational motions? These occur in various frequencies, and the low frequency ones are particularly well organised in the form of waves (in contrast to a solid, the high frequency waves in a liquid are damped). Actually, these low frequency waves are nothing but sound waves. Many years ago, Brillouin predicted that these waves would act like a diffraction grating immersed in the medium (solid or liquid) and that light falling on the medium would be diffracted by these waves. In addition, the light would be frequency-shifted on account of Doppler effect. Remember, sound waves propagate; therefore, when light waves bounce off them, they (the light waves) can get Doppler-shifted.

In short, though the analysis given above is different, Brillouin scattering is nothing but Raman scattering from the acoustic waves in the medium – this comes out when a proper quantum mechanical theory of scattering from the translational motions is worked out. The Russians, Landsberg and Mandel'shtam were actually looking for Brillouin scattering from quartz. They did not find this type of Raman scattering but something slightly different (see Box 5.2).

Box 5.2 This box offers a primer on Raman scattering by crystals. In a crystal, the atoms are arranged in a regular lattice. Consider a simple diatomic chain as in (a). This chain has many modes of normal vibration, two of which are illustrated in (b). Each mode is characterised by an angular frequency $\omega(= 2\pi\nu)$ and a wavevector $q(= 2\pi/\lambda$, where λ is the wavelength). In one of the two modes shown in (b), the adjacent atoms move in the same direction while in the other mode, they do so in opposite directions. The former is called an *acoustic* mode and the latter is called an *optic* mode.

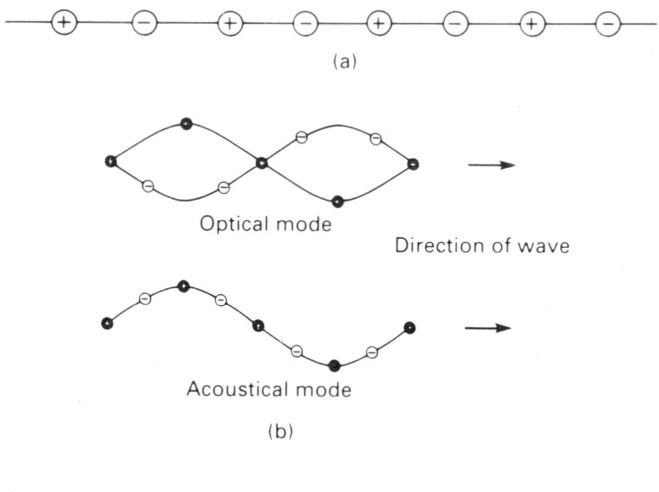

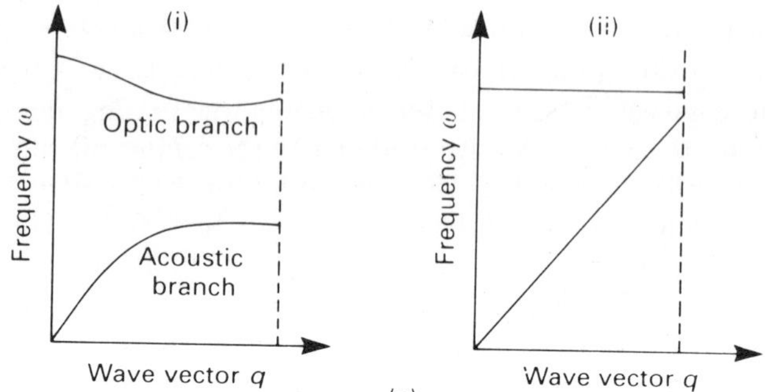

(c)

If a plot of ω versus q is made for the two modes, it would be as in c(i). If all modes had the same velocity, the curves would be as in c(ii). In general, the velocities differ which means that c(i) is closer to reality. Remember the splitting of white light into several colours by a prism? That is because the velocity of light in glass is different for different frequencies. The curves in c(i) are therefore referred to as *dispersion curves* (for lattice vibrational waves).

Raman scattering can pick up acoustic and optic modes of long wavelength (low q). The q is selected by choosing the scattering angle. Raman scattering by the acoustic modes is also called Brillouin scattering. Because ω varies with q for the acoustic modes, the scattered photon has different frequencies at different scattering angles, even though the frequency of the incident photon remains the same. In the case of Raman scattering by optic modes, this is not so.

Landsberg and Mandel'shtam were looking for Brillouin scattering in quartz (i.e., Raman scattering from the acoustic mode). Instead they found Raman scattering by the optic mode. In the thirties, the study of Brillouin scattering by crystals picked up. Many of Raman's students worked in this area.

Incidentally, Brillouin scattering from liquids and crystals was actively studied by Raman's students in the thirties.

Is there no "classical" way of describing the Raman effect? There is, but not an entirely satisfactory one. Essentially, we go back to the induced dipole picture I introduced earlier but with one change. We now say that the atom is a dipole capable of spontaneously oscillating at various frequencies natural to it. For the sake of simplicity, let us say there is only one such frequency ν_0. What is the physical meaning of all this? Figure 5.8 helps us to understand. This shows the energy levels of the atom. Between these levels, many transitions are possible.

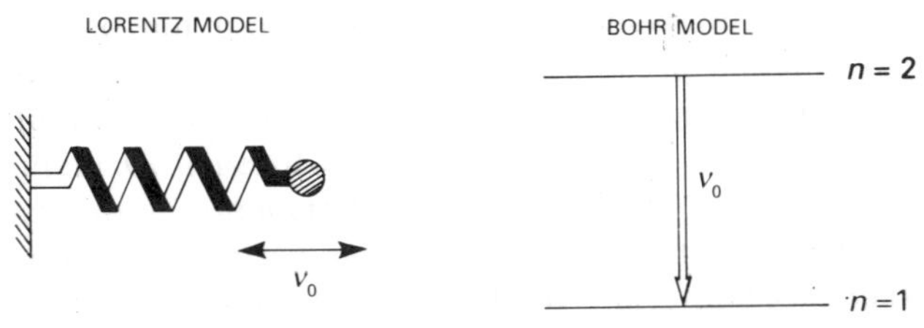

Fig.5.8 The Lorentz model once again, but correlated now to the Bohr model for the simplified case where the Bohr atom has only two levels. How this picture is useful in giving a "classical" explanation of the Raman effect is discussed in the text.

When we represent the atom by a dipole capable of vibrating at various frequencies, we choose these frequencies to correspond to the *transition* frequencies. And when we say that the dipole has only one frequency, it means that there is only one transition frequency, i.e., the atom has only two levels.

The dipole model of the atom was first introduced by Lorentz at the turn of the century. It must be taken with a pinch of salt. If the atom were an oscillating dipole (as is claimed) then would not the atom radiate? And would it not soon exhaust all its energy? These are inconvenient questions. As I have explained in *The Quantum Revolution: Part* I, the dipole picture was a desperate attempt to model the atom for describing its optical properties. Problems increased when the Bohr model of the atom entered the picture for one now had two models – the dipole model which worked for some problems and the Bohr model which worked for others. These difficulties disappeared only when quantum mechanics made its appearance, followed by QED.

Keep all this in mind. Now let us go back to the dipole picture. As before, we suppose that radiation of frequency ω falls on the atom. The atom now acquires an induced dipole moment. The polarisation induced is given by

$$P = \alpha E = P_0 + P_1, \quad \text{say.} \tag{5.9}$$

Here,

$$E = E_0 \cos \omega_0 t, \quad \alpha = \alpha_0 + \alpha_1 \cos(\omega t + \phi) \tag{5.10}$$

Using this in (5.9) we then get

$$P_0 = \alpha_0 E_0 \cos \omega_0 t \tag{5.11a}$$

$$\begin{aligned} P_1 &= (1/2)[\alpha_1 E_0 \cos\{(\omega_0 - \omega)t - \phi\} \\ &\quad + \alpha_1 E_0 \cos\{(\omega_0 + \omega)t + \phi\}]. \end{aligned} \tag{5.11b}$$

The polarisation P_0 contributes to Rayleigh scattering, while P_1 produces Raman scattering; the two terms in (5.11b) result in Stokes and anti-Stokes scattering respectively.

What is the status of Raman scattering today? As Raman correctly foresaw, it has emerged as a tool used extensively by chemists and biologists. Essentially it has become an analytical tool. And the power of the tool has increased immensely after the advent of the laser. It is no exaggeration to say that several thousands of papers have been written reporting applications of the Raman effect. Indeed, there is now an international conference on Raman scattering which takes place every three years. In 1978, the conference came to Bangalore in commemoration of the golden jubilee of the discovery of the Raman effect. Many of the people associated with the glorious Calcutta days had passed away but Bhagavantam was still alive and he recalled those thrilling events to an eager international audience. To many an old timer, it was a nostalgic moment.

6 *Raman At The Indian Institute Of Science*

The Indian Institute of Science in Bangalore is one of the great academic institutions of the country. It was founded in 1911 by Jamshed Tata, the pioneer industrialist who gave India its first steel plant and after whom the city of Jamshedpur (in Bihar) is named. Tata wanted India to be strong and powerful. Industries make a country strong, which was why he established a steel plant in Bihar and power stations in the western part of the country. Tata was convinced that large scale introduction of science and technology alone would solve India's problems. In this, his thinking was rather like that of Mahendra Lal Sircar. But Tata felt that factories and power plants alone would not do. He wanted in addition, centres of advanced learning which would train engineers and scientists, and that was how the institute in Bangalore came to be set up.

The Indian Institute of Science (I.I.Sc.) was started as a centre for applied sciences, with departments devoted to Chemistry and Electrical Engineering. The Director of the Institute was, in those days, always an Englishman. He had a fabulous house to live in and drew the unimaginable salary of Rs.3000 per month. Unlike now, the rupee went a long way and so this was indeed a princely salary. Most of the Professors and senior staff members were also Englishmen. I am not sure if the Institute really came up to old Tata's dreams but it certainly provided nice and comfortable jobs to many.

The Institute was governed by a Council (even today there is a Council), and in the late twenties, Raman was appointed a member. At the time, Dr. Martin Forster was the Director. Dr. Forster was due to retire in early 1933, and so the question came up about a successor. It was the practice to appoint a Committee to look for a suitable person. One such Committee was formed on this occasion also, and as usual it went around looking for suitable persons in England. At this stage, Lord Rutherford was also consulted and he told the Committee that it was wasting its time looking in England when there was an excellent candidate in India, namely Raman. Raman was then offered the prestigious job of the Directorship, and he stepped into it in April 1933, the first Indian to do so.

Raman found the atmosphere at the Institute dull and lacking in dynamism. Besides, the quality of the research work done at the Institute was quite ordinary. Naturally, all this was not to Raman's liking. So he initiated a series of steps which would revitalise the Institute. Unfortunately, this was like stirring a hornet's nest. And you know what hornets do when disturbed. Well, that's what happened in this case also. Raman was attacked and so badly, that he had to step down from the Director's post within a few years; that was the only way he could avoid outright dismissal! All this might sound unbelievable but unfortunately, there does exist such a dark chapter in the history of science in modern India. Wait till you hear more about it!

The Institute had many incompetent staff members and this included several Englishmen. Already there was a resentment amongst some of them about serving under a dark-skinned man and to top it all this dark man was after them with a hatchet because he would not tolerate lethargy and mediocrity. Those affected ganged up together and brought all kinds of charges against Raman. So strong was the feeling against Raman that some of these opponents even wrote nasty letters to Rutherford complaining about Raman. The latter was shocked by the indecent extent to which these opponents were ready to go.

Trouble between Raman and his opponents started almost from the very moment he stepped into the Institute but the last straw was his effort to get Max Born appointed as Professor. This part of the story is so important that I cannot afford to skip over it.

I am sure you have heard of Max Born before. You will find more about him in Part I of the trilogy *The Quantum Revolution.* Max Born was born in Germany but being a Jew, he found the going tough when Hitler came to power in the early thirties. So, like many others, he left Germany seeking a new homeland. His first stop was Cambridge where Rutherford offered him brief shelter. At that time, Born received a letter from Raman requesting that he (Born) recommend the names of young scientists seeking jobs. Raman was aware of what was going on in Europe, that brilliant scientists were running away from Germany and that many of them were on the lookout for jobs. Why not bring some of them to India and thereby raise the standard of research and teaching in the country? This was an excellent idea, especially when you consider the names in Raman's list. Anyway, to get back to the story, Born wrote back saying that he could not make recommendations because he knew nothing about India. Raman then shot back: "Why don't *you* come and spend, say six months and find out what India is like?" When Born showed this letter to Rutherford, the latter said, "This is your chance. Go to Bangalore; anyway, the salary is better than what we offer in Cambridge." (Would you believe that! Those days were different.)

And so Born came to India. He liked Bangalore and the Institute, and Raman for his part liked Born very much. He felt, and rightly so, that Born would be a great asset to the Institute. As Director, he therefore mooted a proposal to have Born appointed as a Professor of Theoretical Physics. A huge howl then went up about irregularities etc. And one Englishman went so far as to describe Born as a second-rate German who was not wanted by his own country. The irony is that Aston, the person who so attacked Born, had earlier stayed with Born for some time soon after he arrived from England, till he could find a house. Born was so shaken by Aston's behaviour that he actually cried.

The story ended sadly. Born went back to England. Raman faced an enquiry and was found "guilty". Eventually, he stepped down from the Director's post. For a brief period he toyed with the idea of leaving the country. With a Nobel Prize in his pocket, he would not have had much problem getting a job abroad. But he dropped the idea and continued to stay on as a Professor of Physics at a much lesser salary than he drew as Director.

Raman stayed at the Institute from 1933 to 1948. For the first five years of this period, he was the Director. He was engaged in research from day one, and this brings us to the scientific work that Raman did at the Institute.

It is not possible to describe here the many problems that Raman explored but a few points need mention. Firstly, the Institute had no physics department when Raman took over. So his first job was to *create* one, and in fact this was one of the factors that contributed to his unpopularity. Incidentally, today the Physics department of the Institute is one of the best in the country, enjoying enviable international reputation. Secondly, Raman groomed many students, several of whom later achieved eminence. The name of Professor G.N. Ramachandran, F.R.S. serves as a good example of the many who were so shaped. Professor S. Ramaseshan, whom I have quoted earlier, is another.

One must also remember that when Homi Bhabha came to India in 1939, from England, he believed it would be a short holiday. He was forced to stay on, on account of the outbreak of the Second

World War, and it was Raman who gave him "shelter" and nursed him for a few years. Just think of Bhabha's monumental contribution to this country (see the companion volume *Bhabha and His Magnificent Obsessions*) and you will immediately realise how much India would have lost if Bhabha had got frustrated here in India and returned to England.

It was also during this period that Raman founded the Indian Academy of Sciences. Today it is considered a great honour to be elected a Fellow of the Academy. Raman also established scientific journals, like the *Proceedings of the Academy* on the lines of the journals published by the Royal Society. The dozen or so journals published by the Academy today can trace their origins to these *Proceedings*.

Perhaps you are beginning to think that Raman was preoccupied more with organising science than in doing research. Not true! Actually, many successful scientists do turn to science management (as it is called) in their later years but not so in the case of Raman. He passionately believed in research and did his best to remain active in research until the very end – something very rare indeed. If he did a bit of science management, it was always *in addition* to research.

And now for a few words about something that Raman did at I.I.Sc., which ranks second only in importance to the Raman effect; this is the Raman–Nath theory. Raman loved waves, and the problem which this theory dealt with had light waves as well as sound waves. What more could he ask for?! I am sure you are puzzled by all these cryptic remarks; let me not keep you in suspense any longer.

I should start with Brillouin scattering. I have already mentioned this earlier but some of that is worth repeating. Consider a solid or a liquid. The atoms are restless and moving all the time. The motion is complicated but can be analysed mathematically, provided one makes some suitable and justifiable approximations. When one does this, the apparently chaotic motion of the atoms turns out to be due to a superposition of orderly waves of different frequencies and therefore of different wavelengths. Remember, these are *spontaneously* generated waves in the medium. They arise on account of the thermal agitation of the atoms. The higher the temperature, the more the number of such waves.

Among these waves, those of the longest wavelength are no different from sound waves, i.e., waves which we can create artificially in the medium by passing sound from a source. This is an important point. Now what exactly is a sound wave? It consists of regions of compression and rarefaction. In other words, when a sound wave passes through a medium it produces a periodic variation of density in the medium. Brillouin said that this periodic variation of density is as if a diffraction grating is immersed in the medium. So, when light is incident on the medium, it should be diffracted. Using simple optics, it is easy to describe this diffraction.

Let λ^* be the wavelength of the sound wave. This means that the density variations can be interpreted as equivalent to a diffraction grating with a spacing λ^*. If now, light of wavelength λ is incident on the medium, then there would be diffraction, the diffracted beam emerging at an angle θ given by

$$\sin\theta = (\lambda/\lambda^*). \tag{6.1}$$

This type of light deflection is referred to as Brillouin scattering.

Brillouin considered spontaneously excited sound waves, and as I told you earlier, Landsberg and Mandel'shtam were in fact looking for such diffraction effects in quartz when they stumbled on the Raman effect. In the early thirties, Debye and Sears in America, and Lucas and Biquard in Europe attempted to verify whether light would in fact be diffracted by sound waves. And they did this by *deliberately* imposing a sound wave as shown in Fig. 6.1.

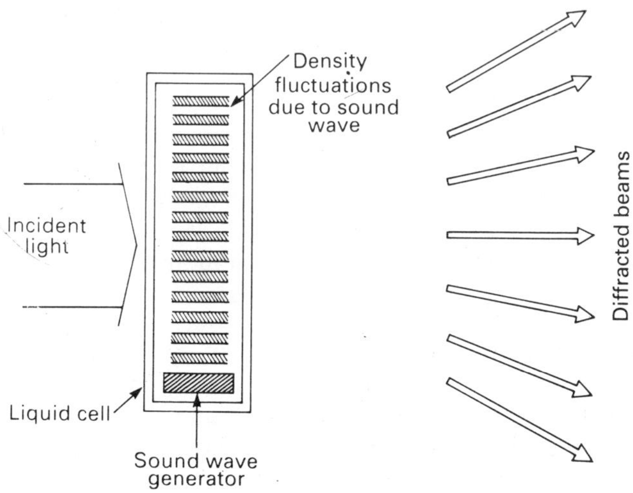

Fig.6.1 Formation of light diffraction bands due to periodic density variations created by a sound wave.

They found that there were several diffraction bands at angles θ_n, given by

$$\sin\theta_n = \pm(n\lambda/\lambda^*) \tag{6.2}$$

where n is an integer. This again is elementary optics and only a mild extension of formula (6.1).

Everything fine so far. Trouble started when people measured the *intensity* of the bands of various orders and how these intensities varied with the conditions of the experiment. Here Brillouin's theory failed completely. For example, when the amplitude of the impressed sound wave was varied (keeping the frequency constant), it was found that some lines became more intense while others became feeble. Raman refers to this as the wandering of the intensities. Why did such a thing happen? No one could explain satisfactorily, and this was the problem.

Raman's attention to the problem was drawn somewhat accidentally. A student named Parthasarathi was doing experiments on sound propagation in liquids and trying to study the velocity of sound in various liquids. (Try to figure out how the study of Brillouin scattering can lead to the velocity of sound.) One day, Parthasarathi gave a seminar on his work during which he mentioned the intensity problem. Raman immediately knew what was wrong with the existing analysis. Let me now quote Nagendra Nath (shortly I shall introduce him) about what happened.

> Professor went to the board and said that the theory should be developed in a different way. A sound wave creates compressions and rarefactions. A light beam would be slowed down in the region of compression and it would move faster in the region of rarefaction, and so a plane wavefront would become corrugated like a zinc sheet used for building purposes. Professor said that an analysis of this corrugated wavefront would explain the unexplained results.

And that analysis job was given to young Nagendra Nath who came up with a preliminary answer the very next morning!

Who was this Nath? He was a young man of about twenty from Bihar, who had been admitted to a course on electrical engineering at the Institute. To Nath, Raman was a hero and he desperately wanted to join the physics department. But his basic degree was in mathematics, and so he lacked the qualification to enter that department. But Nath would not give up. He felt that if somehow he could meet Raman personally, he could make a case and get in. So he sought an interview with Raman with the plea of becoming a research scholar. Let us now hear Nath himself:

> After a thorough interview lasting till the evening he [Raman] stood up and patted me and said that I could join the department as his research scholar. I then told him my plight that I had already secured admission in the Electrical Technology Department. He appeared to be annoyed but then said I had to give up the admission in the E.T. Department. I said that I would very gladly do so.

So that is how Nath became Raman's student. Amongst Raman's students, Nath was perhaps the most gifted mathematically. Max Born had a lot of praise for him, and it is no mean achievement to earn praise from a tough German master like Born.

Once Nath came up with a preliminary explanation based on Raman's ideas, the two worked together and out of this was born the Raman–Nath theory. Let us now concentrate on the underlying physics. It centres around what is called a *phase grating*. To understand what this grating does, consider a 100 metre race. At the start, all the competitors are lined up. During the race, each runner stays in his own track or lane but on account of variations in speed, the runners arrive at the finishing line at different times. Something similar happens in a phase grating. In other words, as the light passes through the grating some parts of the beam travel faster and some parts travel slower. But how can light travel at different speeds like runners in a race? Is not the speed of light a constant? Yes, but that applies to light travelling in vacuum. When light travels in a medium, its speed is determined by the refractive index of the medium. The latter depends, among other things, on the density. If the density varies periodically on account of the sound wave, there are alternate slow and fast tracks for light. The net result is that when light emerges from the "grating" (simulated by the sound wave), the phase of the light wave is not uniform across the wavefront; rather, it alternates, leading to what is called a corrugated wavefront. This is what Nath compares to zinc sheet.

The occurrence of corrugation is illustrated in Fig. 6.2. How they arise can be understood with the help of Fig. 6.3. Here we have a plane wave incident on the grating which can be expressed as

$$\psi(x, y) = e^{ikx}. \qquad (6.3)$$

If the grating did not exist, the form of the wave amplitude would continue to be given by this expression. However, due to the grating, the form of the wave at the exit end x_0 now becomes

$$\psi(x_0, y) = T(y)e^{ikx_0}. \tag{6.4}$$

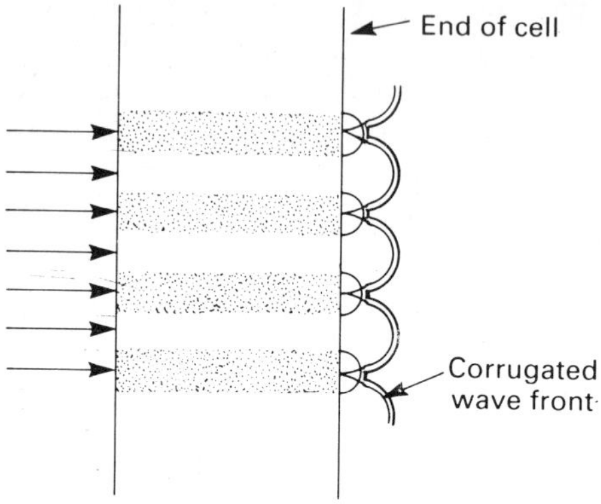

Fig.6.2 Formation of a corrugated wavefront due to the propagation of a light beam through regions of high and low refractive indices. The semicircles on the right show wavelets emerging from liquid cell from regions of compression and rarefaction. The corrugated wave is the envelope of these wavelets.

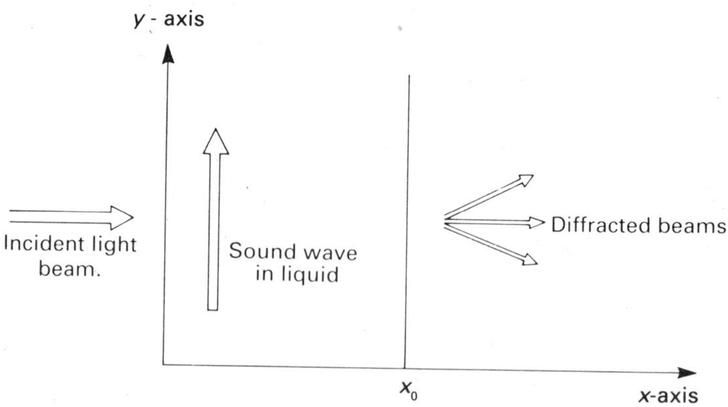

Fig.6.3 Reference framework for describing the action of the phase grating.

It turns out that $T(y)$ has the form

$$T(y) = \exp[i\gamma(y)]. \tag{6.5}$$

Therefore, the wave is now described by

$$\psi(x_0, y) = e^{i[kx_0 + \gamma(y)]}. \tag{6.6}$$

The quantity $\exp[i\gamma(y)]$ describes the phase of the wave, and it varies periodically across the grating (since the refractive index n varies periodically in the y-direction). In short, density variations produced by the impressed sound wave cause a *phase modulation* of the light wave going through – hence the name phase grating.

The idea of the phase grating was nothing new. As early as the nineteenth century, Lord Rayleigh had discussed it in connection with some problem. Raman had read about it. When Parthasarathi was talking, he suddenly recalled it in a flash and saw in it the solution to the problem. Detailed work done later in collaboration with Nath confirmed this.

I have given here a very simplified description of the Raman–Nath theory. You should not get the impression it is all very trivial. Apart from the crucial idea of the phase grating itself, the mathematics underlying the problem is quite challenging. Basically, one is trying to describe the light wave as it is passing through a medium wherein the refractive index is varying. In the Raman–Nath problem this variation is periodic but in general it could be more complicated. What Raman and Nath did was to set up the basic equation whose solution gives the desired description of the propagating wave.

Solving the Raman–Nath equation is a challenge in itself. Years after Raman and Nath did their work, Mike Berry in England did a whole thesis on the subject and even wrote a book.

You might perhaps be wondering what is the great achievement of this theory. Figure 6.4 should provide the answer. You see here a series of patterns, (a) through (x). These are the diffraction patterns observed as the sound frequency is kept fixed but something else, say the amplitude of the sound wave, is varied. Consider, say, pattern (e). We see here a central line which corresponds to the undiffracted (incident) beam and two side bands corresponding to diffraction in the first order. The second-order bands are also faintly visible. Focus on the first-order bands and see how the intensity varies from pattern to pattern. You will notice that while at first the intensity increases, later it decreases, dropping to almost zero

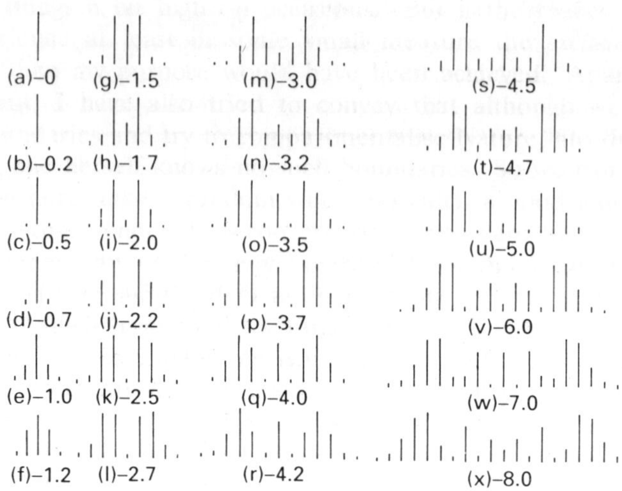

Fig.6.4 Variations in Raman–Nath diffraction patterns obtained by varying the sound amplitude. Observe how the intensity corresponding to a given order varies. This is referred to by Raman as wandering of the intensity.

in pattern (p). It then picks up again. The central band intensity also varies. It is this peculiar intensity variation that Brillouin failed to explain, and which Raman and Nath were able to for the first time.

The Raman–Nath theory is pretty but that is not the reason why it is remembered today. For a theory in physics to survive, be recalled, etc., it must be useful. In what way is the theory of Raman and Nath useful? Essentially in practical applications.

The subject of applications is huge and unfortunately I cannot go into it here. But I must give you some idea of how this usefulness comes about. Before this, I must first mention one fact. When light passes through a phase grating, it is diffracted; and the Raman–Nath theory helps us to understand the intensity of the diffracted waves. Now when going through the grating, the phase of the wave gets altered. What about the *frequency* of the light wave? Does that remain the same or does it change? If so how?

You must have heard of the *Doppler effect*. The standard example is that of a railway engine blowing its whistle and speeding past a station. For an observer on the platform, the sound appears to vary in frequency, even though the engine whistle is producing sound at a single frequency. For the observer, however, the sound is coming from a source which is itself moving; and that is what causes the change of frequency as registered by the observer. This is the essence of the Doppler effect.

In this case also there is a Doppler effect because the sound wave in the liquid (say) is not a frozen wave; rather, it varies in time. The net result is that there is also a Doppler shift, the nature of which is illustrated in Fig.6.5. The essential point is that though the light might go in with frequency ν, it comes out with a different frequency whose value depends on the band. The frequency shift occurring in Brillouin scattering (see also previous chapter) is conveniently visualised in terms of the Doppler effect.

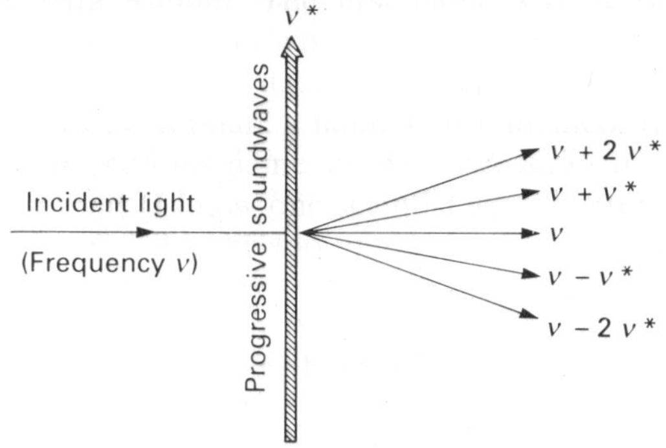

Fig.6.5 Doppler shifts produced in a Raman–Nath experiment.

In the usual lab experiments, (as in the original experiment of Debye and Sears for example), the sound wave impressed is of a single frequency and is produced by a quartz oscillator immersed in the liquid. Suppose that instead of a quartz oscillator we have, say, a quartz microphone and we speak into that. The frequency of the impressed sound wave would depend on what frequencies we feed into the mike. Thanks to the Doppler shift, the diffracted light wave would carry an imprint of the input frequencies. In other words, we can *modulate* light beams with speech and use light as a carrier for sound. To put it differently, the Raman–Nath effect if one might call it that, opens up the possibility of a kind of optical communication. No doubt I have grossly oversimplified things, but that is the hub of it. Optical transmission of speech and data has a lot of advantages and is expected to become the thing of the future. It is in this context that the theory of Raman and Nath has acquired great importance. Meanwhile, the modulation technique has found use in many military applications. By the way, thc cell used for light modulation is often called the Bragg cell possibly because of the Bragg "reflection" taking place. It would be more appropriate to call it the Raman–Nath cell. I might also mention that this cell has been used in a crucial experiment to test some of the philosophical foundations of quantum mechanics – for more information, see the companian volume *The Quantum Revolution: Part* III.

November 1948. Raman attained the age of sixty, and it was time to retire – not from work or research but from I.I.Sc. What did Raman do after retirement? The next chapter tells you the story.

7 *The Final Years*

Raman's retired life stretched from 1948 to 1970 – for twenty-two years. What did he do during that long period? He created a research institute and kept himself busy with research. In addition, he ran the Academy and edited the journals *Current Science* and *The Proceedings of the Academy* .

The institute Raman founded is called the Raman Research Institute. It is located about a kilometre away from the I.I.Sc. The land came as a gift from the then Maharajah of Mysore.

Institutes cannot just spring from nowhere. To equip them and to run them, money is needed. At the time he retired, the Government of India offered a grant to Raman but he refused to accept it. He correctly reasoned that along with Government money come Government controls; and that was something he was not prepared to accept. So Raman raised his own funds. He went up and down the country urging philanthropists, princes (in those days, there were many princes and maharajahs) and wealthy industrialists to contribute. He once described his mission as begging but was not ashamed of it. "Our greatest men were beggars – the Buddha, Sankara and Mahatma Gandhi," he declared.

The donations were not enough. So Raman started a few small-scale industries with the help of some of his old students. The earnings from these he ploughed into the Institute. Even this was not enough. And so he sold some property to raise money. He was the recipient of the Lenin Peace Prize from the Soviet Union (now extinct). This money also he used up for running the Academy and the Institute. As for himself, he was appointed National Professor in 1948. This gave him a modest monthly income with which he managed.

When Raman moved to the new Institute in 1948, it was hardly ready. Professor A. Jayaraman, a former student of Raman recalls:

> For the first year at the Raman Research Institute there was no electricity, but that did not deter Raman from carrying out several beautiful optical experiments with sunlight, a few lenses and a pair of polaroids. He considered a beam of sunlight as the best source and

> in Bangalore there was no shortage of blue sky and bright sun. A manually operated heliostat kept in order by voice communication, produced astonishing results.

One might dismiss Raman's efforts by saying that all this is trivial research. Soon I shall point out that this was not so. But here I would like to make another point which is that unlike all of us who would probably have remained inactive, complaining about lack of this or the lack of that, Raman was determined to do *something* at least. This is a very important lesson for us in India. Nothing is easy here, and compared to the West we are almost always at a disadvantage either in terms of money or equipment or both. If we lose ourselves in comparisons, we would end up feeling miserable and wallow in self-pity.

Raman was made of different stuff. His was a fighting spirit, which he maintained till the very end. In fact he was of the opinion that one does one's best under slightly adverse circumstances. Once at the I.I.Sc. he had given a student a certain spectroscopic problem to work on. One day Raman found the student looking rather dejected. When asked what the matter was, the student replied, "Professor, my competitor in England who is working on the same problem has a 100kw lamp whereas I have only a 10kw one." Raman replied, "Never mind. Put a 100kw brain on it!" True Raman cracked a joke, but his reply also reflects his spirit.

Before I say something about the research that Raman did during this last phase, a few words about how he built up the Institute. There weren't enough staff members simply because of scarcity of funds needed to hire them. But Raman had a gardener with whose help he developed the campus beautifully. You don't have to take my word for it; just go to Bangalore and visit the Institute. You will know what I mean. About Raman's love for his garden, Jayaraman has this to say:

> Raman loved trees, flowers and above all, his rose garden. All the best roses that Bangalore nurseries could supply were bought and planted in his rose garden under his supervision, and he admired them like a child would admire a new toy. He knew the botanical name of every tree in the campus, and he had them planted carefully to maximize the effect of their floral display. Every day he would go round the garden twice, to enjoy, to relax, and to think.

Raman also loved to collect crystals, gems, minerals, rock specimens, shells, stuffed birds, butterflies – anything that displayed colour.

Commenting on Raman's passion for accumulating such objects, Jayaraman observes:

> Raman had an exquisite collection of quartz crystals and quartz family minerals... He was never tired of acquiring quartz crystals, although he would say he would not go in for any more quartz. There was a man from Coimbatore who used to work for the Telegraphic Department. He would bring quartz crystals, small and large, to sell, and Raman's resolution would break down at the sight of the material!

Raman also collected diamonds, which he called the prince of solids. He says:

> I have since the year 1930, been deeply interested in the physical investigation of the diamond. The difficulty of obtaining the material in a form suitable for exact studies has, however, been a serious obstacle to progress. Indeed in the early days, I was reduced to the expedient of borrowing diamond rings from wealthy friends who, though willing to oblige, were slightly apprehensive about the fate of their property.

Bhagavantam once narrated his experience as a student during the days following the discovery of the Raman effect. He was given the job of studying the Raman effect in diamonds, and for this purpose, Raman had borrowed a big piece from some prince. Bhagavantam says:

> He borrowed one big piece (140 carats) for me from one of the then wealthy Maharajahs. The loan was for a limited period and against a security bond for a fairly large sum of money which he had to sign before borrowing it. When he put this rare diamond in my hand for intensive spectroscopic studies in the following 48 hours, he cautioned me to be careful and not lose it and told me how history is replete with stories of professional thieves following great Indian diamonds to stumble on the right opportunity for whisking them away from the owners.

And so young Bhagavantam spent two nights literally near the spectrograph, while the recording was going on! Bhagavantam adds that Raman's passion for diamonds continued all his life. At one stage, almost all his students were working on one aspect or the other of diamond. He says:

> The need for diamonds, in all sizes, shapes and quality, became so

> much that Raman began acquiring them by all methods such as purchasing, borrowing from shops and wealthy owners, buying in auctions and from individuals and so on.

Raman regularly attended the auctions at the Panna diamond mines in Madhya Pradesh, and according to one report, he had as many as 310 diamonds in 1944!

What did Raman do with his various collections? He organised them into a museum, about which Jayaraman comments:

> The specimens were neatly arranged in glass shelves and were lighted appropriately for the best possible viewing. One little room had luminescent minerals and these came to life with brilliant colours when ultraviolet (UV) lights were turned on. This was one of the thrilling experiences for a visitor to the Raman Institute. When this section of the museum was put into operation, Raman must have turned the UV lights on and off a hundred times, each time enjoying the sight like a child.

Raman loved taking visitors through his museum, and in later years he made it a point to take groups of young people through it, keeping them enthralled with descriptions and stories. The museum is there even today but the diamonds are no longer in public view – for obvious reasons of course. I have seen them once many years ago and it was an unbelievable experience.

What kind of science did Raman do in his later years? He did many things of the type people used to do in the last century. One might say this is pointless and a waste of time. Yes, Raman's research was not in the mainstream, if that is the point. In the fifties and the sixties, the fashionable subjects were nuclear physics, particle physics and solid state physics (of a kind). Raman did not have the resources to dabble in these areas. Instead he stuck to what he could afford. Does that mean he did work which was either trivial or irrelevant? The answer is clearly NO. One just has to see how his students fared. He did not have many during this period but they have all given an excellent account of themselves. Two of them, Jayaraman (whom I have quoted a few times) and A.K. Ramdas (son of L.A. Ramdas whom I quoted in an earlier chapter) migrated to America after doing their Ph.D. under Raman. Jayaraman had a distinguished career at the famous Bell Labs, America's premier research centre, while Ramdas still serves as Professor at Purdue University. Both are eminent solid state physicists, and they received their training at the Raman Institute under Raman himself.

One other person whose name must also be mentioned is that of Pancharatnam who, alas, is no more. At Raman's suggestion, he investigated the optics of minerals. You might wonder what is great about this subject. Certainly in the late fifties when Pancharatnam did his work, not many thought it was important. But he systematically explored all the complications that occur when light passes through the minerals. Minerals are often doubly refracting which means that a light beam passing through it splits into two: the ordinary ray and the extraordinary ray. Then there are various effects connected with anisotropic polarisation properties, not to speak of absorption. It is not just that Pancharatnam dealt with a messy problem. He also discovered elegant ways of dealing with such problems and in recent years, these methods have become a part of the resurgence that optics is undergoing, particularly after the advent of lasers and the opening up of the wide field of nonlinear optics. The inspiration for all these studies came from Raman.

Thanks to the Nobel Prize, Raman became an important public personality. He constantly remained in the news not only on account of the various honours that he continued to receive throughout his life (he was, for example, one of the very early recipients of the Bharat Ratna) but also because of his caustic comments on public affairs, especially where they concerned the promotion of science. Naturally, newspapermen loved him!

I should not let you get the impression that Raman craved being in the news and therefore made controversial statements. Far from it. He was *genuinely* disturbed by the spending of large sums of money in the name of science and scientific research, without proper attention to whether the money was being spent usefully and wisely. Fifties and sixties was the period when a large number of labs were being established in the country. Big buildings were being built and all kinds of equipment was being imported. Raman questioned the need for all this.

Money sure is needed to promote science, but good science essentially comes out of well motivated and competent people. In India this obvious fact is often lost sight of. This might sound surprising but unfortunately it is true and Raman took upon himself the painful duty of telling society where it was going wrong.

Understandably, Raman was not liked by many of the big wigs of the day and he was criticised and also attacked in many ways. Raman was deeply hurt and became somewhat of a recluse. Eventually he overcame his depression by taking active interest in young students, particularly those in school. He loved taking them in groups through his Institute, talking to them about science and showing them the museum and the garden.

On the 2nd of October every year, he delivered a lecture called the Gandhi Memorial Lecture (the tradition of having that lecture continues). The last time he gave that lecture was in 1970. That year he spoke on the theory of hearing, and as usual, illustrated his talk with live demonstrations of various kinds. Shortly afterwards, he fell ill.

The end came in the early hours of November 21, 1970. By a special arrangement, his mortal remains were confined to the flames in the Institute campus itself, amidst surroundings he loved and enjoyed. According to his wishes, there were no religious ceremonies. Thousands of people – school children, students, colleagues, friends and others – gathered to pay homage. Raman had passed into history.

No plaque or monument marks that spot – just a solitary tree, precisely as Raman would have wished. Next time you go to Bangalore, visit the Institute, stand before the tree, and reflect for a few moments on what this great son of India has achieved.

8 *Raman – An Overview*

Before I wind up, I would like to look at Raman's personality as a whole. Normally, I don't do this when I describe the work of a scientist but I must make an exception in this case because Raman was much more than just a brilliant scientist.

When I started my scientific career in the mid-fifties, Raman had already retired. At that time, he was better known for his controversies than for his accomplishments. In fact, most people of my generation knew little or *nothing* either about Raman's struggles or his achievements, other than that he had discovered the Raman effect and that he had won the Nobel Prize. There were many Raman stories, several of them intended to show Raman in poor light as a person who knew little of science. We were swept by contemporary scientific events, and took a dim view of someone who talked optics and acoustics – to many of us, that was not science at all; at best it was old hat. To make matters worse, there were people of the older generation, many of whom knew Raman personally and who were totally unabashed in their praise for "Sir C.V." We had little respect for these "old fogeys" as well, and felt quite proud of our disdain for the old man and his contemporaries. Today I feel ashamed to admit all this but alas, it is true. Many of my contemporaries share this feeling.

What made us all so foolish and arrogant? Basically, it was our ignorance – ignorance about who Raman was, what he went through and what he had achieved. We knew next to nothing about the enormous difficulties he had faced at various times as well as what he had achieved *inspite of these adverse circumstances*. For example, none of us knew of his valiant and wise move to get Born appointed at the I.I.Sc. and the treatment that was meted out to him for so trying. I did not know anything about it until a few years ago when I was asked to write a detailed biography of Raman on the occasion of his birth centenary. I might add that in 1988 when I gave a lecture at I.I.Sc. on Raman's stay at the Institute, the facts I revealed were totally unknown to those assembled there. It came as a great shock to them, and the audience included a former Director who expressed amazement about the way Raman had been treated.

A few words now about Raman's approach to science. Raman was prolific in his output, and published several hundreds of papers. Their style is very different from what we are used to today and in fact this is partly responsible for the tendency of many to underrate Raman's scientific merit and calibre. In so misjudging Raman, we fail to see that Raman grew up reading the masters of the last century like Rayleigh, Helmholtz, Faraday and the like. Naturally, his outlook was akin to that of nineteenth century physicists. He continued to remain fascinated by the problems of that era and did not hesitate to investigate the colours of shells, for example, even though it might not have been contemporary fashion. To Raman, science was the highest form of creative art. And the problems he chose often dealt with natural phenomena which had a touch of beauty. He never ceased to be amazed or be struck by things beautiful, and then went on to enquire about the scientific basis of this beauty. In this he was rather like the well-known British physicist G.I. Taylor who described himself as an amateur scientist who worked for his own pleasure.

In his time, Raman was criticised for his views on science policy, management and funding. Those were the days when the country was led to believe that science and technology would help overcome all its problems. True, India needed (and continues to need) a big dose of modern science and technology for solving her problems. Raman never questioned that. He knew only too well the importance of science; after all, had he not been the beneficiary of the dreams of Mahendra Lal Sircar who himself held such views? Raman himself once said, "What India needs is science and more science..." So it was not science that Raman was against but the *way* in which it was sought to be promoted.

Raman's fears have proved true. Today, the public, the press and Parliament are rather disenchanted with the output of our laboratories. Thousands of crores of rupees have been spent but we do not have enough to show for it. On the contrary, the import of technology is going up by leaps and bounds.

It is not that we are just importing technology. What is even more disturbing is that a kind of paralysis has crept into our thinking. For some reason or the other, we have not only developed a deep suspicion for things Indian, but, to make matters worse, developed also a great and uncontrolled craving for things foreign.

Consider, for example, the craze for going abroad. Everybody has it, and scientists are no exception. But desires should be within limits, and scientists of this country should certainly be guided by the examples set by eminent persons like Raman, Bose, Saha and Bhabha. Unfortunately, that is not the case. I can dwell at length on this topic but will not. I will only draw your attention to Raman's refusal to go to England for training and exposure, even though it was supposed to be one of the conditions for appointment as Palit Professor, this despite the fact Raman had never been abroad before.

It is not merely a question of going abroad. We have a rather slavish mentality about working mostly on problems which attract attention abroad. The late Mr. J.R.D. Tata (he is so eminent that I am sure you know about him) once gave an example of this attitude. He was the President of the Family Planning Association of India. He strove hard to promote research on family planning. But he complained in a TV interview that there was much resistance to research on this subject. The interviewer asked why. Mr. Tata replied, "Because there are no international conferences on family planning. Our scientists do not want to work on problems which cannot take them on foreign trips."

Frequent junketing is not the only malady. (I must admit that this is so only with a select few but nevertheless, since this "few" includes important people, the examples they set have an impact.) There is in addition a sad lack of interest in the way science is being shaped and in the way science is receiving attention in schools and colleges. Raman took particular pains to lecture to school and college students wherever he went. Countless are the people who have been inspired by these lectures. How many of our present-day research scientists are willing to lecture to school children?

Raman repeatedly lamented about the neglect of our universities. Today they are in a much poorer shape than in Raman's time and yet how many dare or care to raise their voice?

Most scientists today are content to take a defensive and antiseptic view of things. Their attitude is: "My job is to do 'good' science. I cannot change the system. That is someone else's headache. If things are not satisfactory, that is just too bad but I can't waste my time over it." If we don't wish to face our problems and would like to run away from them, who would solve them for us? Remember, at the height of his humiliation, Raman stayed back; and thanks to that, we have had G.N. Ramachandran and a string of other brilliant scientists. And we should not forget Homi Bhabha who though not a student of Raman, was nevertheless provided "shelter" by Raman during a crucial period. Just think of what Bhabha has meant to the development of science in the country (for more about him, see *Bhabha and His Magnificent Obsessions*).

It is quite common for an Indian scientist today to seek collaboration with someone in Poland (say) rather than with his colleague in his own lab. Theoretical physicists abroad are ready to work with experimentalists in their lab but in India such collaborations are conspicuous by their absence. No doubt there are excuses, but is this the way to build up science in the country? Is it any wonder that everybody thinks of import? When somebody wants to design, he would like a foreign prototype to copy. When somebody wants to do an experiment, he would like to have imported equipment and imported samples. When theorists in India want to explain the results of an experiment, it is almost invariably the results of experiments done abroad. Finally, practically everyone likes to go to foreign conferences to present the results and eventually publish them in foreign journals which most institutions in India cannot afford to buy.

We might patronisingly dismiss Raman's research in later years as being out of tune with contemporary trends. But I think there is something to be said for the spirit of the man who chose to remain independent and not be swept by fashion. Following fashions and keeping up with the Jones' is not always the best thing.

Many years ago, during K.S. Krishnan's period, the National Physical Laboratory started research on solar energy. At that time, oil was dirt cheap and solar energy seemed irrelevant, at least to many other countries. And so, because it was not fashionable, we gave up research in this area. Decades later came the oil crisis and solar energy suddenly became popular. Meanwhile, we had slipped badly behind.

In justification of many of our actions, we often hear the phrase that science is international. True, but if you examine this carefully, it is so only at the surface. Leaving aside all the political and economic implications associated with this question (and there are many) let us look at it from a practical point of view.

When we try to work on a so-called current field, we straightaway face many handicaps. To start with, we must wait until the news reaches us. This could take several months after the discovery or breakthrough is made. And then, if it is experimental work, it would be months and months before we can put our act together. Meanwhile, things move rapidly in the Western world. Lots of things happen because people there can travel frequently (travelling from Holland to Germany is like going from Bombay to Nagpur or even

Pune), they can easily exchange information, using fax and e-mail, speak over the phone, etc. We simply can't afford all this. Eventually our choice narrows down to the following: (i) Get frustrated and join the camp of complainers ("Things are no good in this country."). (ii) Migrate permanently. (iii) Make good contacts and friends abroad and regularly shuttle between the West and India.

Many "successful" scientists have settled for option (iii) which has lead to the joke that they are half in the chair and half in the air! Let us examine these questions dispassionately. Is it really possible for *all* Indian scientists to emigrate? Will the countries of the West be prepared to receive *lakhs* of them? Is it sensible to practice "shuttle science" neglecting research and teaching here? (Believe me, this regularly happens!).

Why do we see such trends today? Because we largely think of ourselves, our achievements and rewards, and seldom of our society or country as a whole. True, for a scientist (like an artist), rewards mean much more than for an average person. But if the quest for rewards leads to massive neglect of national institutions, then it becomes a malady and ceases to be a desirable quality that spurs one on. When it came to national spirit, there were few who could outshine Raman. After all, did he not claim Gandhi as his "dearest friend"?

Today, our scientific community complains bitterly. "Oh we don't get support. Merit is not recognised. Bureaucrats throttle us. Look at what politics has done." Yes, *all* the complaints are valid. But then, should not one stand up and fight against wrongs? Does one just look the other way or go off to other countries? Was that the example set by Mahendra Lal Sircar? Did Raman migrate in his darkest hour? If we don't attempt to speak out and set things right and let whatever has been established decay, what will happen to the students of tomorrow?

Raman realised these ills well before they became serious maladies. He spoke sharply and instead of being thanked for the kind warning, he was belittled. Today one sees how right he was but the sad part is that his message has not sunk deep yet. Two countries which set a much higher regard for self-reliance and building up from grassroots than us are the (former) USSR and China. They have not only the gloss of science to show, but behind it also a strong structure. We have only a facade; behind it, not much to speak about.

Basically, it is a crisis of spirit and nothing else. Don't think I am making an imaginary complaint. If you have read the volume on

Bose in this series, you would perhaps recall what Bose said while concluding the Krishnan Memorial Lecture. Bose wondered why we are a third-rate country, inspite of being blessed with abundant resources, both material and human.

Raman had the answer. He strongly believed, and I agree with him one hundred percent, that it is due to a defeatist feeling that is widely prevalent. In a stirring speech he once declared:

> I would like to tell the young men and women before me not to lose hope and courage. Success can only come to you by courageous devotion to the task lying in front of you and *there is nothing worth in this world that can come without the sweat of our brow*. I can assert without fear of contradiction that the quality of the Indian mind is equal to the quality of any Teutonic, Nordic or Anglo-Saxon mind. What we lack is perhaps courage, what we lack is perhaps driving force which takes one anywhere. We have, I think, developed an inferiority complex. I think what is needed in India today is the destruction of that defeatist spirit. We need a spirit of victory, a spirit that would carry us to our rightful place under the sun, a spirit which will recognize that we, as inheritors of a proud civilization, are entitled to a rightful place on this planet. If that indomitable spirit were to arise, nothing can hold us from achieving our rightful destiny.

Raman spoke these words a quarter of a century ago. They appear more valid and relevant today than they did when first uttered.

We must also remember that Raman firmly rejected the attractions of office, preferring to immerse himself in science. In retirement, he not only refused memberships of various prestigious committees but even an offer of the Vice-Presidentship. "What would I do with that ship?" he is reported to have said. How many of us are prepared to serve the cause of science with such devotion?

In Chapter 3, I quoted Dr. Fermor who predicted that one day Bangalore would possibly make a claim as a city of scientific excellence. Today his prophecy has come true. Thanks to the seed planted by Jamshedji Tata and the nursing of academic excellence by Raman during the crucial years, Bangalore has truly become *the* premier science city of the Nation, including within it not only a dozen or so leading laboratories but also industries ranging from those dealing with aircrafts to computers and space technology. All this has become possible through hard work and deep faith in our inner strength. However, we could have and ought to have done much more.

Raman constantly reminded us about our strength. He is now long gone, and his voice has been replaced by those of doubters. Raman certainly deserves to be remembered for his towering scientific accomplishments. But today, we need to honour him *even more* for his fighting spirit, a spirit he had in abundant measure all his life, and which we, as a nation, sorely lack at present.

Suggestions for further reading

1. Venkataraman, G. *Journey Into Light: Life and Science of C.V. Raman*, Penguin Books India, 1994.

2. Jayaraman, A. *C.V. Raman*, Affiliated East West Press: New Delhi, 1992.

Index

Note: Page numbers in brackets refer to illustrations